GETTING THE PICTURE

A GUIDE TO CATV AND THE NEW ELECTRONIC MEDIA

OTHER IEEE PRESS BOOKS

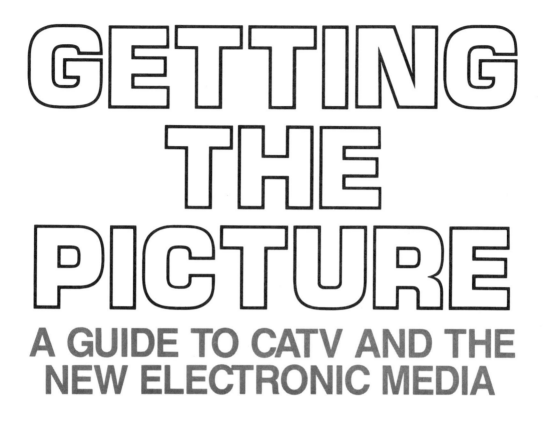

GETTING THE PICTURE

A GUIDE TO CATV AND THE NEW ELECTRONIC MEDIA

STEPHEN B. WEINSTEIN

Published under the sponsorship of the
IEEE Consumer Electronics Society.

IEEE PRESS

The Institute of Electrical and Electronics Engineers, Inc., New York.

Copyright © 1986 by
THE INSTITUTE OF ELECTRICAL AND ELECTRONICS ENGINEERS, INC.
345 East 47th Street, New York, NY 10017-2394
All rights reserved.

PRINTED IN THE UNITED STATES OF AMERICA

IEEE Order Number: PC01891

Library of Congress Cataloging-in-Publication Data

Weinstein, Stephen B.
  Getting the picture.

  ''Published under the sponsorship of the IEEE Consumer Electronics
Society.''
  Includes index.
  1. Cable television. 2. Data transmission systems. 3. Telecommunica-
tion systems. I. IEEE Consumer Electronics Society. II. Title.
TK6675.W45 1986   384.55′4   86-7275

ISBN 0-87942-197-5

ACKNOWLEDGMENT

This book could not have been completed without the support, encouragement, and advice of many individuals. I am especially grateful to Beth Bogie of Warner Amex Cable Communications, David Carlson of Stern Telecommunications Corporation, Geof Gates of Cox Cable Communications, Joe Garodnick of Phasecom Corporation, Eric Addeo of Bell Communications Research, and Bob Braudy of Logica, Inc. for information and criticism. I thank all of the knowledgeable and considerate individuals in the video and electronics industries who provided data and illustrative materials. All opinions expressed in the book, and any errors, are, of course, entirely my responsibility. The bulk of the writing was done while I was employed at the American Express Company, and I am indebted to the Company for its support, and to its employees Anne Graham and Aileen Haley for their essential secretarial and word processing services. I owe many thanks to Reed Crone, Managing Editor of the IEEE PRESS, for his good advice and continual encouragement. Finally, I am deeply grateful to my wife, Judith, and my children, Brant and Anna, for their unwavering patience and support throughout the long period of time over which the book was written.

v

DEDICATION

To Judith, with love.

Contents

Preface

The U.S. cable industry is a fast-growing collection of electronic programming, distribution, and equipment businesses whose products have captured the interest and a great deal of the recreational time of the television viewing public. The basic product is television by wire. The emergence of cable as an important broadcasting medium with a high penetration of the potential market—now about 45 percent of U.S. television homes and expected to reach 50 percent soon—has come after many false starts and the investment of billions of dollars. Distribution systems franchised by local governments have cropped up everywhere, and most (not all) metropolitan areas should be wired by the late 1980's. With the passing of the financial difficulties of the "big build" era and the resolution of major regulatory uncertainties in the Cable Communications Policy Act of 1984, cable is likely to settle in as a stable element of the electronic communication environment.

The era of cable development is, however, just beginning. As a widely available communications medium, cable is so new that few of the serious questions about its technical capabilities, competitive viability, social impacts, and eventual role in the general communications infrastructure have been answered. Some commentators, although fewer than in earlier years, feel that cable, multi-channeled and sometimes interactive, will overwhelm broadcasting and other electronic and print media. Others think that reception of programming directly from satellites, or movies on videocassettes, will take away cable's audience. Social and moral critics draw attention to transmission monopolies, individual privacy, cultural programming, "electronic democracy," and the delivery of R- or X-rated movies. Telephone executives are concerned about competition from cable in basic voice and data services. City governments worry about cable systems reneging on contracts and overcharging consumers, while cable operators worry about cities gouging them with excessive taxes and service demands. Cable operators, faced with tougher financial conditions than they anticipated, have scaled back their services, sought rate increases, and asked whether investment in two-way cable and other prerequisites for new services is justified at all.

There are many possible answers to these concerns, and no rational person will want to make up his or her mind without considering the facts. This book is an attempt to put forward a body of facts concerning the history, technology, programming, regulation, and competitive stance of cable distribution systems. It emphasizes technology and in that sense is a technical book, but no technical background, other than a general appreciation for recent events in electronics and communications, should be necessary to understand it.

The major thrust of this book is to describe cable itself, but after that it is to describe, and compare with cable, several important competitive media. ''Free'' television broadcasting is omitted, even though it could be regarded as CATV's biggest competitor, because it is not one of the new media that have become significant in recent years, or may become significant in the near future. The telephone network *is* included because of cable's challenge to it in data communication services, and because the telephone network is evolving new capabilities for video transport.

For future development and success, it is my opinion that cable will have to defend and enlarge its revenue base in video services through technical improvements, pay-per-view marketing of individual presentations, and distribution of programming intended for relatively specialized audiences, as well as branch out into new revenue-generating communications services. It will face serious competition in every area. It is becoming clearer day by day that a diversity of electronic delivery media will be sustained by a combination of market demand, government policy, and technical good sense. Cable will not have a monopoly in any service, but will instead share a large market with terrestrial broadcast services, direct satellite broadcasting, microwave communication systems, private cable systems, videocassette recorders, and telephone-line interactive services. It may eventually face competition from a vastly enhanced ''telephone network'' consisting of switched optical fiber transmission facilities extending out to almost every home, but it is possible that video services providers and telephone companies will find a way to share this very expensive upgrading of the nation's communications network. Cable operators may one day abandon their proprietary coaxial cables and become service providers via a ubiquitous public light-wave network.

These alternative communications media, which vary in capacity, interactive capability, and geographic reach but are competitive with today's CATV systems in some or all services areas, are concisely described in this book, with the overall intent being to define the place of cable systems within the larger communications picture. Perhaps the reader will come to agree that services will be increasingly transferable from one delivery medium to another, and that this will encourage the integration of local cable systems into a great interconnected wide-band communications network spanning the nation and eventually the world, whether or not the universal optical fiber network becomes a reality.

That, of course, is the future. Cable television, as it exists today in the United States, is delivered by more or less isolated local systems used largely for one-way distribution of video programming to month-by-month subscribers. The distribution systems are built on the long-established technology of coaxial cable, which is an electrical conduit capable of carrying as many as 70 television channels or any equivalent load of communications traffic. Cable television is sold for its combination of good reception and uninterrupted movies (on the "pay" channels), and "pipeline TV" has become the principal source of entertainment for many households despite its frequent failure to meet customer expectations for programming, signal, and service quality.

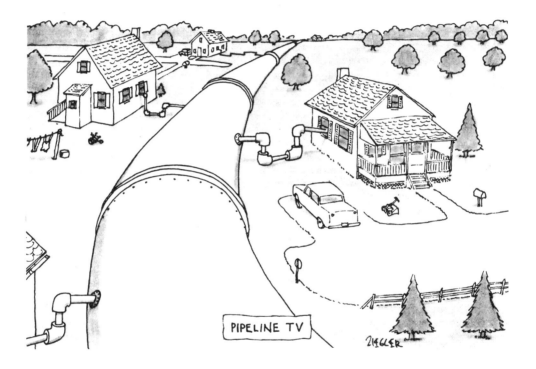

PIPELINE TV

Drawing by Ziegler; © 1982 The New Yorker Magazine.

The construction and operation of cable systems has been carried out under the authority of local governments, frequently after heated and occasionally sordid franchise competitions. At times when the profit potential of the medium appeared large, a variety of companies entered the industry, ready and willing to invest hundreds of millions of dollars to acquire existing cable systems, build new big-city franchises, or become program distributors via communications satellite. Publishers have invested to protect their long-range interests as information suppliers, and broadcasters to keep their markets. A relatively small group of multi-system operators and major program distributors appears to be solidifying as the long-term industry structure.

The industry has lost its high-profit glamour, but the penetration of cable into a majority of television homes in the United States seems assured. Observers now discuss *what* cable should do and will be able to do rather than whether it *will* be available. There are differences of opinion on specifics, but general agreement that the industry must improve its service quality, consolidate geographically, and clearly differentiate itself from broadcast television if it is to be financially viable in the long run.

This book covers a broad territory, and cannot claim to do more than provide the beginnings of an explanation of complex technologies and controversial economic and public-interest issues. Industry structure is constantly changing, and the reader should understand that the descriptions given were current when this book went to press. But because of its broad coverage of distribution technologies, services, and policy issues, this book should contribute to an informed public opinion about the potential of cable and other electronic communications media and how best to use them.

Stephen B. Weinstein

1 Growth of the Industry

Cable television was born out of the desire of people living in remote areas to receive good quality broadcast television. Local entrepreneurs placed large television antennas on high points of the local terrain, and used coaxial cables and signal amplifiers to distribute the received signals to subscribers. A very early system, and perhaps the first in the United States, was established in 1948 in Mahanoy City, Pennsylvania, by John Walson, who built it to help sell television sets at his appliance store. Astoria, Oregon, also lays claim to the first distribution network. Known as community antenna television, these systems carried a few broadcast signals and were sometimes community-owned. In 1962, there were less than 800 cable systems in the United States, with a total of 850,000 subscribers.

Most of the cable industry's growth until the late 1960's was in community antenna systems in towns which either had little available off-the-air television or had reception difficulties because of hilly terrain. The number of signals available was frequently increased by microwave relay of programming from metropolitan areas, as illustrated in Fig. 1.1, but the total number of video channels offered to subscribers was conventionally no more than 12, covering the standard VHF band. It is surprising that as late as 1978, 70 percent of all U.S. cable systems still offered 12 or fewer channels.

Broadcasters were not too concerned about the pickup, without compensation, of their signals for distribution in locations with poor reception, but they began to become concerned when cable systems were constructed even in good reception areas as a means of increasing the choice available to subscribers. Signal importation was encouraged by the increased availability in the 1950's and 1960's of microwave technology and frequencies. As an important side effect, the use of microwave transmission systems in cable businesses allowed the FCC to establish its authority over cable, which could be viewed as an appendage to the microwave services which the FCC was already regulating. Twenty-channel systems became practical in the 1960's with the introduction of channel converters (see Chapter 2), further enhancing cable's ability to deliver a wide choice of programs.

The importation of distant signals for carriage on cable systems was opposed by the broadcast industry, by now an implacable foe of cable, as unfair competition to local broadcasting, and in 1966 the FCC prohibited all further signal importation into the 100 largest markets, where a reasonable

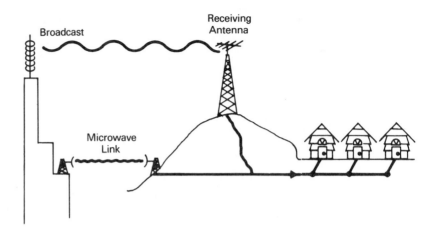

Fig. 1.1: Community antenna television, with additional microwave feed.

selection of programming was presumably already available. The regulatory authority of the FCC, challenged by Southwestern Cable Company, was affirmed by the U.S. Supreme Court in 1968.

In 1968, the FCC, sympathetic to the position of the broadcasters, actually forbade new cable construction in the 100 largest markets. This attitude had changed somewhat by 1972, when the FCC's new rules governing cable were issued and the ban was lifted. However, the rules were still highly restrictive. For example, cable systems could not show films that were between two and eight years old. The general attitude of the FCC at that time, and not only in regard to cable, was described by Commissioner Glen O. Robinson in 1976 and quoted (unfavorably) by a later Chairman, Charles D. Ferris: "If it moves, regulate it; if it doesn't move, kick it—and when it moves, regulate it" [4]. (See the "For Further Reading" section at the end of each chapter for numbered references.)

At about this time, the concept of the "wired nation" began to take hold in the minds of officials, technicians, and entrepreneurs. Expectations of swelling income streams dazzled the cable entrepreneurs, who began to compete heavily for new franchises, which have been and continue to be issued by local or municipal authorities. Cable companies, most of them small and highly leveraged (debt-financed), promised the world to city administrations, as they did again on an even larger scale ten years later. Many of them soon faced disaster from a combination of rising interest rates, excessive franchise terms, poor penetration, and stagnant subscriber rates. The public did not show much interest. The largest operator in 1973, Teleprompter, took a $37 million write-off that year.

Amidst all the turmoil and despite all the restrictions, pay cable—the basis for the rapid expansion of the industry which began a few years later—was launched in November 1972 by Home Box Office (HBO), a small subsidiary of Sterling Manhattan Cable, which was largely under the control of Time, Inc. "Pay service" is programming for which the subscriber pays an

Fig. 1.2: *RCA's Satcom 1, the first major carrier of programming for cable systems, launched December 1975. (Courtesy of RCA Americom.)*

extra fee, in addition to the monthly fee for "basic" cable service. It had actually been started in California in 1963, but an intensive compaign by movie-theater owners, including a referendum on a California constitutional amendment prohibiting anything other than "free TV," killed it as soon as it began. The FCC's constraints on pay cable movies and sports were not lifted until a federal court of appeals compelled them to be lifted in 1977.

HBO had a microwave distribution capability in New York City, and supplied programming on tape to its first out-of-town customer, John Walson's Service Electric Cable TV System in Wilkes-Barre, Pennsylvania, for distribution to 365 subscribers. HBO was taken over completely by Time, Inc. in September of 1973, and by the end of 1973 was serving about 8000 subscribers on 14 systems. An expansion to nearly 60,000 subscribers on 42 systems by the end of 1974 was made possible by a microwave network in the Northeast, replacing the tape distribution system. HBO was heavily bankrolled by its parent, and did not make money until 1977.

The major growth of HBO and of the cable industry started in late 1975, when HBO began distribution of programming by the RCA communications satellite Satcom 1 (Fig. 1.2). This brilliant and daring business innovation made it possible to distribute programming at moderate cost to virtually any location in the United States and Canada. A number of satellites dedicated largely to cable service were subsequently launched. Satcom 3R, placed in orbit in November 1981, replaced Satcom 3 which was lost in space soon after its launching in late 1979, and services on Satcom 1, approaching the end of its useful life, were transferred to Satcom 3R in January 1982. Satcom 4 was launched at that time, and Western Union's Westar 5 went up in June 1982. Hughes Satellite Communications launched Galaxy 1, a higher-powered satellite, in June 1983. The constellation of satellites carrying video programming continues to expand.

Satellite distribution was almost immediately successful, although at first only the large multiple-system operations were able to install earth stations because of an FCC rule requiring use of 9-meter (almost 30 feet) receiving dishes, which cost as much as $100,000. The FCC, beginning a remarkable era of relaxation of its constraints on cable operations, soon reduced this restriction to 4.5 meters (15 feet, Fig. 1.3), which dropped the earth station price to as low as $15,000. Technical progress in television receive-only (TVRO) technology has made it possible to achieve top-quality performance in stations of relatively small size and low cost, and the FCC's willingness to relax unnecessarily strict technical standards is only one of the ways it has tried to give the cable industry the incentive and freedom to grow quickly.

Now virtually every cable system desiring satellite-fed programming can afford to set up one or more satellite-receiving stations, in addition to broadcast, microwave, and local studio feeds (Fig. 1.4). This technical and marketing advantage, together with the increasingly permissive attitude of federal regulatory authorities, has fueled an enormous expansion in the cable industry and in its expectations for the future. Between 1979 and 1983, the industry spent $6.7 billion for new and rebuilt plants. The total number of subscribers in the roughly 6000 U.S. cable systems grew to 38 million in 1985—44 percent of television households, as shown in Figs. 1.5 and 1.6.

Much of the prosperity of the cable industry depends on viewing of the non-broadcast programs, both pay and advertiser-supported, which are carried by almost all cable systems. The cable programming share of the total television-viewing audience grew to about 16 percent in 1984, and an apparent downturn in cable services audience the year before was reversed, with cable-originated program services and satellite-delivered ''superstations'' becoming the fastest-growing segments in television programming in the first nine months of 1984. If a 1985 Federal District Court ruling

overturning the "must carry" rules under which a cable system must dedicate channels to local broadcast stations is upheld, satellite-distributed cable programming could expand its audience share considerably.

Table 1.1 lists the top ten cable operators, by number of subscribers, in 1984. All of them are multiple system operators (MSO's), owning many individual cable systems. Many of these operators are also heavily involved in program production and distribution (see Chapter 3), in ancillary services (see Chapter 4), and in other sectors of the communications, entertainment, and broadcasting industries.

Although Table 1.1 shows very large revenues—the New York Times reported an industry total of $6.1 billion for 1983 compared with $27 million in 1975—cable profitability went down in the early 1980's, and the prospects for future profitability are only moderately good. The "churn," or cancellation rate, of pay subscriptions is estimated to be about 40 percent per year, suggesting a fundamental consumer dissatisfaction.

There are other problems, too. Theft of services nationally is thought to be about $500 million per year, with the New York State Cable Television Association estimating that between 10 and 50 percent of the homes that can receive cable are using it illegally. Higher copyright royalties are being paid, and a more demanding urban customer is increasing the servicing

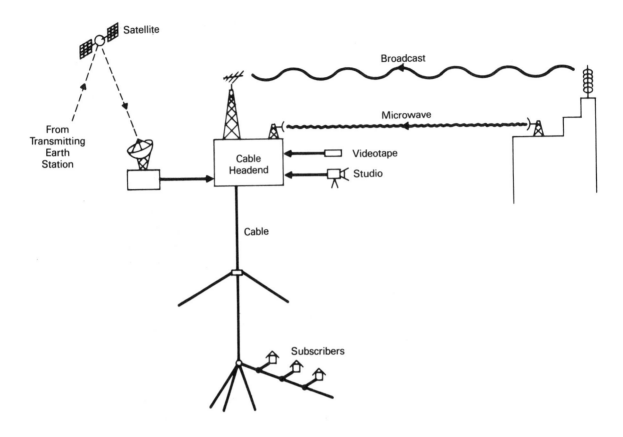

Satellite

From Transmitting Earth Station

Broadcast

Microwave

Cable Headend

Videotape

Studio

Cable

Subscribers

Fig. 1.4: Elements of a modern cable system with satellite, broadcast, microwave, videotape, and studio feeds.

expenses of cable-system operation. Satellite-delivered basic (non-pay) program services are increasing their fees to cable operators as a result of their own losses, and many program services have folded or been merged with others.

Perhaps the largest problem is the "urban builds"—the enormously expensive new cable systems in the country's largest cities—which cable companies agreed to build under contracts which were later seen as impossible to fulfill at a profit. Wiring costs in cities have been very high—one 1983 estimate was $900 to $1200 per subscriber—and have been frequently underestimated. Warner Amex Cable Communications, perhaps the most active builder among the MSO's, spent over $80 million instead of the projected $47 million to wire Pittsburgh, and some of the later builds were expected to cost hundreds of millions of dollars. Low subscription prices and lavish public-access facilities were promised to cities, based on high market penetrations which now appear difficult to realize. As Drew Lewis, Chairman and Chief Executive Officer of Warner Amex, said in 1984, "We just promised too much, and now we find out that to break even, we can't live up to those promises" [5]. In the mid 1980's,

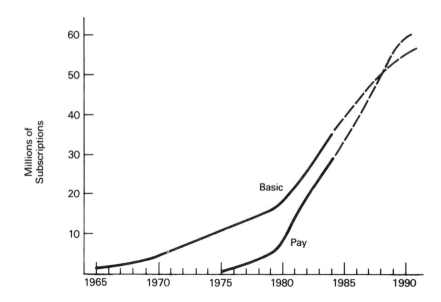

Fig. 1.5: The growth of cable television, with future projections. (Source: NCTA "Cable television developments, Feb. 1985" Paul Kagan Assoc. Inc., Dec. 1984.)

Fig. 1.6: Penetration of cable in the United States by early 1985. "Pay" is number of subscribers with one or more pay subscriptions. (Source: NCTA, Apr. 1985; Paul Kagan Associates, Apr. 1985.)

renegotiation of franchise agreements became a high priority for cable operators with big construction obligations.

The highest priority for cable operators will always be the maintenance of an adequate revenue stream from residential subscribers. This is determined by three factors: basic penetration, or percentage of homes passed which subscribe; pay penetration, or number of pay-channel subscriptions sold to the customer base; and the rates charged for basic and pay services. Cable operators try to provide a mix of services which will maximize subscriptions, and have fought against efforts of local authorities to exercise control over the prices and content of optional cable services, finally winning a large measure of deregulatory freedom in the Cable Communications Policy Act of 1984, described in Chapter 5.

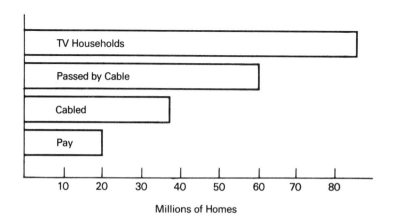

Table 1.1: The largest multiple system operators (MSO's).*

Company	Parent Company	End-1984 Subscribers	End-1984 Monthly Cable Revenues $Millions	Programming and Service Interests
Tele-Communications (TCI)		3,500,000	60	Taft-TCC Programs Cabletime Magazine
American Television & Communications (ATC)	Time, Inc.	2,500,000	50	Home Box Office (HBO) Cinemax USA Network (1/3) Tri-Star Pictures (with Columbia and CBS) EvenTelevision others
Group W Cable	Westinghouse Electric (prior to sale to TCI, ATC, and others)	2,000,000	39	Nashville Network Home Theater Network EvenTelevision
Cox Cable Communications	Cox Communications	1,600,000	32.8	INDAX (past) CommLine (past) Pay per view Rainbow
Storer Cable Communications	Storer Communications	1,500,000	28.4	New England Sports Network Magnicom Systems
Warner Amex Cable Communications	50% American Express 50% Warner Communications (prior to Warner buyout of American Express)	1,200,000	20.3	Showtime/The Movie Channel (w/Viacom) MTV, VH-1, Nickelodeon (prior to anticipated sale to Viacom)
Times Mirror Cable	Times Mirror Company	1,000,000	21.3	Videotex America
Continental Cablevision	75% private 25% Dow Jones	950,000	19.5	
Newhouse Broadcasting	S.I. Newhouse & Sons	900,000	17.8	
Viacom Cable	Viacom International	800,000	14.8	Showtime/The Movie Channel (w/Warner and Warner Amex) Lifetime (w/Hearst/ABC Video)

*Sources: *Cable TV Investor*, Paul Kagan Assoc. Inc., Dec. 3, 1983; *New York Times*, Aug. 23, 1985 (quoting Kagan data); *CableVision*, Oct. 29, 1984.

For the newer systems with at least 35 channels, the average per-subscriber revenue in 1984 was in the range of $20–25 per month, or $240–300 per year. Many cable operators are developing ancillary services (see Chapter 4), directed to both residential and business users, to supplement this income stream, although early enthusiasm has considerably waned. The cable industry still has great hopes for pay-per-view programming, for which

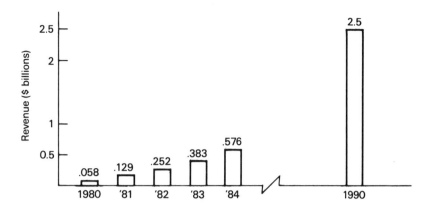

Fig. 1.7: Cable advertising revenues, past and anticipated. (Sources: New York Times, Mar. 23, 1982; CableVision, May 7, 1984; Advertising Age, Jan. 21, 1985.)

payment is made by the program or even by the minute. One industry spokeman forecast $5 billion in pay-per-view revenues by 1990. Many in the industry are counting on advertising revenues (Fig. 1.7), although CATV, with 1984 advertising revenues of a little under $600 million, has a long way to go to catch up with the broadcasting industry's $16 billion.

"Regional clustering" is a cost-containment and business-strengthening step which has been pursued by many cable MSO's. The major advantages to the MSO are reduction of the high administrative cost of operating widely scattered systems, and consolidation of audiences to attract advertising.

In one concept of clustering, 7 to 25 systems use centralized billing, marketing, service, and technical staffs, which give them the economic advantages of size while enabling them to be bigger in any single market than many competing technologies. Clustering can be either among systems owned by different operators, as when a big-city franchise combines some operations with neighboring suburban systems, or by a single MSO which attempts to buy a number of systems in a metropolitan area. Trading of systems to accomplish this has become a common practice.

Creating a large audience for advertising—and larger advertising revenues for cable operators—is possible even without a full-scale business coordination. Many "ad interconnect" service organizations for distribution of advertising to groups of cable operators have been created in recent years. The Bay Area Interconnect, a joint venture of Gill Cable and Viacom, tying together the systems of ten different California operators, expected $3 million in advertising revenues in 1985, and $6 million in 1990. Programming of regional interest, such as local sports, can also enjoy a larger audience and thus be produced and distributed more economically, and sharing of satellite, microwave, and studio facilities can further reduce costs.

Industry Structure

Although the cable industry has so far been described as if it consisted solely of cable systems and operators, it actually has four major components:

cable systems, program distributors (or simply *programmers), program producers,* and *equipment suppliers.* Many of the programmers, such as HBO, are owned by or associated with cable MSO's. Program producers— mostly film studios, are increasingly involved financially and operationally with program distributors.

The relationships among producers, distributors, and cable operators are critical in the competition for the consumer's dollar and to the health of the industry as a whole. The CATV program distribution industry partly cajoled and partly forced its way into the movie studios, which in 1983 made $1.07 billion, or 24 percent of their annual revenue, from pay television and videocassettes. Steps have been taken by distributors individually or in joint venture, as described in Chapter 3, to corner the movie production of one or another major studio or at least develop a steady supply of movies for their pay channels. Several distributors are sponsoring film productions on their own.

But the continuing general unavailability to cable of first runs of major motion pictures has kept CATV from reaching its full potential. With cable interests so deeply involved in motion picture production, this is no longer a question of arbitrary studio preference for theatrical distribution, but a simple economic decision to maximize revenues. This situation will change only when cable implements pay-per-view in simple and cost-effective ways.

Part of the tangle of industry relationships, and the process of consolidation, can be seen in a cursory examination of several of the larger cable MSO's. The industry structure is still in flux, and the examples given here may have changed when this is read. Time Inc. began with small cable operations and achieved an early and spectacular success as a program distributor (HBO). It bought American Television and Communications (ATC) in the late 1970's and built it, through a series of acquisitions, into one of the largest MSO's. Manhattan Cablevision is its best known system. It was a partner with TCI in the buyouts of Warner Amex and Group W Cable in 1985 (described later). ATC also owns three multipoint distribution systems (MDS, see Chapter 9) and had interests in two subscription television stations (STV, Chapter 7) until selling them in 1983. ATC has often swapped isolated cable systems with other operators.

Time, Inc. has a history of program production and distribution, although its enthusiasm for production may have waned. In early 1985 it owned the Cinemax movie channel in addition to HBO, and had a one-third interest, along with MCA and Paramount, in the USA Network, which carries sports and variety programming. A subsidiary of HBO was involved in a joint venture for film production, called "Tri-Star Pictures," with Columbia Pictures and CBS, and HBO had an agreement with Columbia guaranteeing its right to carry Columbia films. HBO had a joint venture with Orion Pictures Corporation to develop programming, and a similar agreement on film rights.

HBO also arranged with Twentieth Century Fox Film Company and

Universal Pictures for nonexclusive rights to their films. A subsidiary of HBO made a licensing agreement with Silver Screen Partners, which gave HBO exclusive pay-TV rights to their productions. HBO further set up its own production company, HBO Premiere Films. Expanding internationally, HBO was in a joint venture with Goldcrest Films and Television, Columbia Pictures, and Twentieth Century Fox to create a pay-television distributor for British cable systems, and an extension to the European continent. On the pay-per-view front, Time, Inc. was in a joint venture, "EvenTelevision," with Tele-Communications, Inc., Group W Cable, and Caesar's World, although it had not yet made a full commitment to pay-per-view television.

Time, Inc. has had smaller interests in the equipment sector. It set up a joint venture with Toshiba Corporation for development of multi-unit addressable decoders, a technology which discourages signal theft and can support pay-per-view, and another with M/A-COM Linkabit for signal-encryption equipment. It abandoned earlier investments in teletext development and in a cable magazine.

Tele-Communications Inc. (TCI), the largest MSO, is far more concentrated in cable operations. It is a well-financed company which developed almost entirely through acquisitions and partnerships. In 1984, TCI bought the Pittsburgh system, with 170,000 homes passed, from financially pressed Warner Amex, and in 1985 joined Time, Inc. in purchasing the rest of Warner Amex's cable systems and, later, Group W Cable.

TCI had a minority interest in a pay service, Spotlight, which was dissolved in 1983. At the time of writing its only participation in program production and distribution was through a joint venture with Taft Cablevision Associates which operates Black Entertainment Television, an advertiser-supported cable network, and film distribution (not by satellite) via its National Telefilm Associates subsidiary.

TCI published *Cabletime Magazine*, a program guide. It was not involved in equipment manufacturing, but had a large user interest in addressable decoders (see Chapter 2).

Warner Communications, until 1984 the owner of the Atari electronic game company, began its activity in cable in 1968–69 with several modest acquisitions. Its major system was in Columbus, Ohio, until American Express, a large financial services company, bought a half interest in 1979 for $175 million. The resulting joint venture, Warner Amex Cable Communications, won several major franchise competitions, partly on the basis of its "QUBE" interactive communication capabilities, almost abandoned in 1985 because the audience did not warrant the cost and an adequate advertising revenue stream did not develop. QUBE was capable of supporting pay-per-view TV, but was not extensively used for this purpose. This heavy commitment to expensive new construction resulted in serious business losses, typified by a $99 million after-tax loss in 1983.

Warner Amex, whose serious financial difficulties led to piecemeal sales of cable systems and to Warner's buyout of American Express's half-share for $450 million in 1985, emphasized programming as much as cable system operation during its years of independence. Its programming affiliate in its last days of independence was MTV Networks, Inc., which went public in 1984 but was still largely owned by Warner Communications and the American Express Company. MTV Networks distributed the highly successful Music Television (MTV) channel; Nickelodeon, a high quality children's programming service; and Video Hits One (VH-1), a popular music channel aimed at adults. Warner Amex's The Movie Channel, overshadowed by HBO, as were most other pay services, merged, in 1983 with Showtime to form a more competitive pay service. Warner Amex retained a minority interest. Warner Communications and Warner Amex's interests in both programming services were sold to Viacom International in 1985. Warner Amex occasionally showed interest in developing ancillary services, and made some progress with burglar and fire alarm security systems until the subsidiary doing this work was sold at the end of 1984.

Westinghouse Broadcasting, a long-time owner of television and radio stations that had only small investments in cable, acquired Teleprompter Corporation in 1982 for $646 million and renamed it Group W Cable. It became the third largest MSO, operating 150 systems in 34 states. At the end of 1985, Westinghouse agreed to sell Group W Cable for $1.6 billion to a consortium of cable operators including TCI and ATC.

Westinghouse was also active in programming services. It sold its interest in Showtime in 1982, but continued to distribute Home Theater Network, a pay service; and The Nashville Network, an ad-supported country music service produced by Opryland USA, Inc. Westinghouse also owned and distributed regional sports networks in the Washington, D.C., and Seattle/Tacoma areas. It has been a participant in the EvenTelevision pay-per-view venture. Westinghouse also owned and distributed the Muzak background music service, and operates Group W Productions, a cable program marketing organization. Group W Satellite Communications, another subsidiary, owned a substantial number of satellite transponders and offered these, and engineering services, to the cable and broadcasting industries.

Like other companies, Westinghouse was active in interconnects and clustering. It was a partner with Viacom in the Pacific Northwest Interconnect, which sells advertising on a number of Seattle/Tacoma area cable systems. ATC and Group W Cable swapped systems in 1983 to enhance the clustering of each MSO.

Cox Communications, owner of a number of television and radio stations and the fourth largest MSO, has been successful through growth by acquisition and has been a sponsor of interactive services and technologies. Its INDAX system (see Chapter 2), dropped in 1984 for lack of consumer interest, was intended as the vehicle for a combination of television, information, home banking, and shopping services. Cox has also pursued "bypass" communications technologies circumventing telephone company facilities. One of its partners was Chase Manhattan Bank. Its CommLine business communications service, first offered in Omaha, Nebraska, did not grow rapidly, and was closed down at the end of 1985. At one time, it was seen as the cable industry's strongest effort in non-entertainment communications. Cox has also entered the cellular mobile telephone business.

Cox has not been heavily involved in program production and distribution, although it has had an interest in Rainbow Programming Services, which distributes Bravo, American Movie Classics, and The Playboy Channel. A previous interest in the Spotlight movie channel was sold to Viacom and Warner Communications in 1983. Cox has been active in program syndication through Bing Crosby Studios and Telerep.

These examples of a few of the largest MSO's throw light on the participants, turmoil, and multiple interests in the industry. The cross-couplings with the entertainment, publishing, and communications industries are evident, and would be even more so if further examples were given. The cable industry is a complicated web of alliances and cross-ownerships which has not yet settled down to a stable pattern.

One of the more interesting evolutionary themes is that of the relationship between the cable and telephone industries. Although there is mutual distrust over bypass and pole attachment issues (see Chapter 5), a long-term cooperative relationship is possible. One obvious arrangement is the cable-telco hybrid interactive (two-way) system (see Chapter 2), which is increasingly viewed as the most practical way to implement interactive services. Furthermore, telephone companies, although generally forbidden to operate CATV systems or offer programs to cable viewers, can build and own cable facilities and lease them to separate CATV operators.

In Washington, D.C., for example, District Cablevision, Inc. contracted in 1984 with the Chesapeake and Potomac Telephone Company to build and own its distribution network, contingent on FCC approval. In a slightly different arrangement, Wisconsin Bell Telephone was authorized by the FCC to build a broad-band distribution service in Brookfield, Wisconsin, a Milwaukee suburb, with channels to be leased to unaffiliated entities including the CATV franchise holder, Telenational. Earlier cable operator objections to leasing rather than owning distribution facilities appear to have lessened, and the main question now is cost. To a cable operator, who is more a programming retailer than a communications builder, it can make good sense to leave construction and maintenance of the communication facilities to others, particularly if franchise renewal is clearly independent of network ownership. Telephone companies may in the future become even more dependent on CATV leasing as they introduce wide-band optical fiber into their local loop (subscriber telephone line) plant, and the decision to do so could hinge on CATV usage.

Some telephone regulators and telephone company managements also see cable systems as an important, if temporary, part of the communications infrastructure which could be integrated with the telephone network in the near term. "Cross use of facilities would lower the cost of phone service," according to a 1983 statement of Gerald Irwin, a spokesman for the Michigan Public Service Commission. Bulk transmission for business customers and communications services on hybrid interactive systems are the most obvious cross-use examples, as described in Chapter 4.

The Future of the Industry

This turbulent, entrepreneurial, and sometimes mystifying cable industry has many problems to overcome and serious challenges from the alternative distribution media. But when the favorable signs are examined—the multichannel capacity of cable systems with relatively simple technology; the good signal quality regardless of geography and obstructions; the potential to offer a variety of services to meet the needs of diverse and specialized audiences and user groups; the relaxation of regulatory constraints; and the financial strength of the dominant MSO's and programmers—it is hard to predict anything but a bright future.

Coaxial cable systems are likely to enjoy many years of success as the principal distribution medium for video programming, and to have small but significant businesses in other communications-based services. The utility of the cable medium is being expanded through regional clustering and interconnections with other communication facilities. Even if coaxial cable distribution systems become obsolete, which could happen if a public fiber optic network were built out to most residences and businesses, cable operators would be able to continue as service packagers and providers on the new delivery system, and program distributors could only benefit from higher quality subscriber transmission facilities.

The cable industry has had a difficult time because of its growing pains, its incomplete understanding of the markets for its services, and the vacillating attitudes of regulatory authorities. With maturity close at hand, it is even possible, as industry analyst Dennis Leibowitz predicted in 1984, that profits will pour in during the late 1980's as capital costs decline and almost everyone subscribes. Industry consolidation and an improved financial picture should bring stability and new opportunities for development of CATV into the more exciting medium which was envisioned in the 1960's.

For Further Reading

[1] "History of cable television," National Cable Television Association (NCTA), Washington, D.C., 1984.
[2] T.E. Baldwin and D.S. McVoy, *Cable Communications*. Englewood Cliffs, NJ: Prentice-Hall, 1983.
[3] "The emergence of pay cable television," Technology + Economics, Inc., Cambridge,

MA, vol. 1, pg. 7, July 1980, (report prepared for the National Telecommunications and Information Administration).

[4] "FCC news," rep. 15953, Oct. 2, 1980 (an FCC publication).

[5] "Cable operators take a bruising," *The New York Times*, March 4, 1984.

[6] T. Whiteside, "Onward and upward with the arts: Cable 1," *The New Yorker*, May 20, 1985. Parts 2 and 3 were published on May 27 and June 3, respectively.

2 How It Works

The basic operating unit of the cable industry is a "cable system," consisting largely of a municipal-scale coaxial cable network to distribute radio-frequency television signals to subscribers' television sets. The network may also use optical fiber and microwave transmission.

As Fig. 2.1 shows, a cable system resembles an upside-down tree, with signals carried from the cable headend to distribution hubs, from which they are sent through trunk, feeder, and drop cables to subscribers' homes. This architecture, sometimes called "tree and branch," is appropriate for the primary commercial function of the cable system, which is to broadcast entertainment programming in quantity and with high received quality at all subscriber locations.

An alternative switched star architecture (Fig. 2.2), in which video signals are switched, just like telephone calls, in the hub of a star to individual subscriber lines, is favored in some European countries and Japan. It offers opportunities for broader selection of program materials, but is considerably more expensive to build. The cable operator is able to control, in the hub, the signal flow of one or a few channels to individual subscribers. A smaller subscriber hub, compatible with tree and branch networks, has been offered in the United States by Times Fiber Cable.

CATV wiring can be strung from telephone poles for a fee paid to the telephone or power company, or buried underground (Fig. 2.3), according to local ordinances and requirements. In larger cities, underground ducts used for a variety of communications wiring will usually accommodate a few new cables. Existing ductwork is often used in apartment and office buildings. The wiring is carefully planned on detail maps (Fig. 2.4), which remain important to maintenance and upgrading after completion of the system.

Capital costs—for buildings and land, headend equipment, distribution plant, and subscriber equipment—are substantial. Expressed as a cost per mile for new construction, they are $7500 to $15,000 per mile for aerial construction, and $15,000 to over $100,000 per mile for underground construction. It is of critical importance to the economics of cable construction to sign up a large percentage of the residences passed by cable, and not to lose cheaper-to-wire apartment complexes to private cable and other competitors.

Commercial cable systems are often grouped into three major categories:

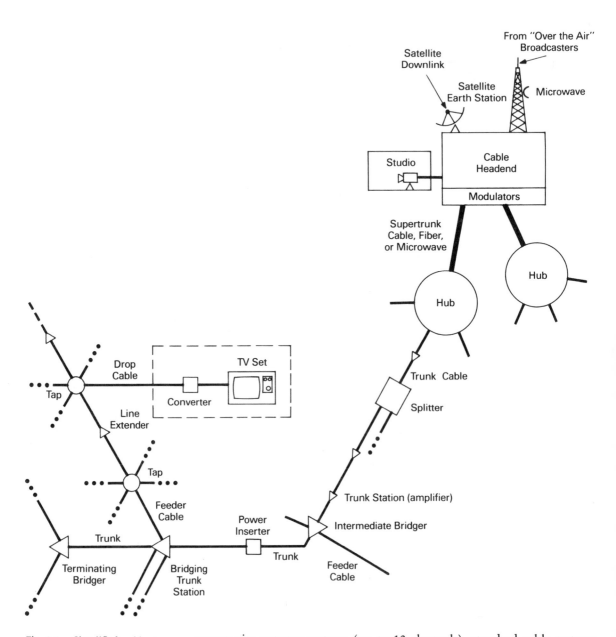

Fig. 2.1: *Simplified architecture of a cable system. A feeder cable may supply 100 to 200 subscribers. Regional hubs serve different parts of a metropolitan area or simply split up a larger subscriber base into a number of service groups of 10,000–40,000 subscribers each.*

community antenna systems (up to 12 channels), standard cable systems (13–35 channels), and wide-band cable systems (more than 35 channels). Wide-band systems, which are relatively new on the American scene, may use either one or several parallel cables. New cable systems usually consist of multiple separate networks, one for residential users and one or more for institutional and commercial users, such as government offices, hospitals, schools, banks, and manufacturers. These separate networks can be

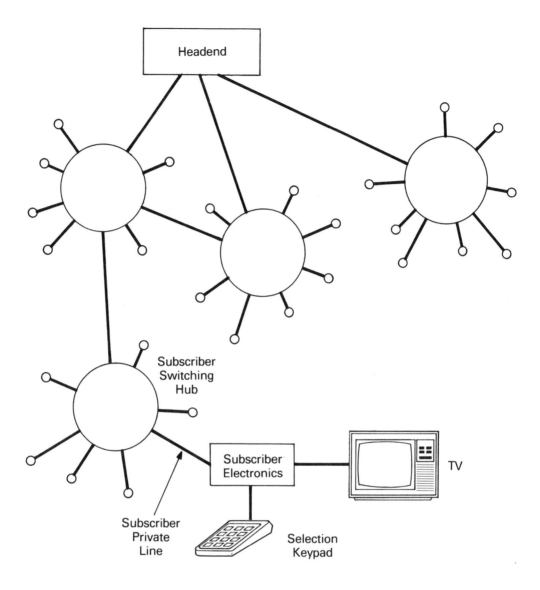

Headend

Subscriber
Switching
Hub

Subscriber
Electronics

TV

Subscriber
Private
Line

Selection
Keypad

Fig. 2.2: *The switched star cable network architecture, rarely used in the U.S. Each subscriber has a program selection device which communicates over a private cable or fiber line with the local distribution hub. A greater or lesser degree of interconnection among the stars is possible.*

interconnected so that they can exchange communications traffic and potentially link residential subscribers with institutional and business users. The institutional and commercial networks may not necessarily have the upside-down tree appearance of the residential network; special communications connections among users can be made as needed and ordered.

In this chapter, we will first introduce the major transmission media which are used to carry signals in cable systems—coaxial cable, microwave, and optical fiber—and then move down through the delivery network from

Fig. 2.3(a): Installing aerial cable.

communications satellite to subscriber television set, describing each part of the system along the way. Most of the discussion will be readily intelligible to the non-technically trained reader, but several sections, particularly on interactive systems, call for some technical background. The glossary at the end of the book should be helpful to all readers. The non-technical reader can skip over the technical details without any serious loss of understanding of the general discussion.

Fig. 2.3(b): *Installing underground cable.*

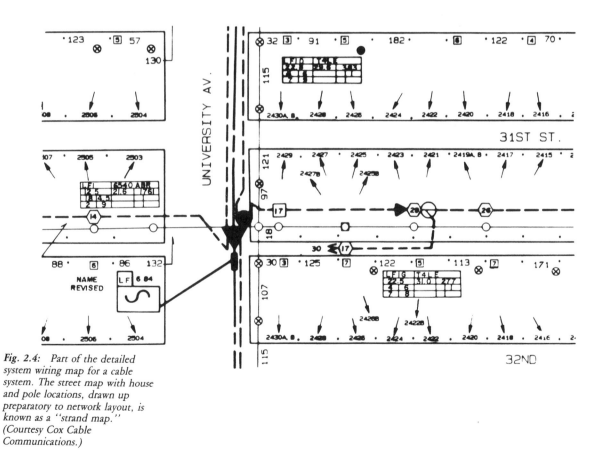

Fig. 2.4: *Part of the detailed system wiring map for a cable system. The street map with house and pole locations, drawn up preparatory to network layout, is known as a "strand map." (Courtesy Cox Cable Communications.)*

Coaxial Cable, Microwave, and Optical Fiber

Coaxial cable (Fig. 2.5) is a transmission medium for electromagnetic signals, consisting of a center conductor and a concentric conducting sheath separated by a solid or gaseous material which is non-conducting. Sometimes the cable is pressurized to maintain its shape and provide a mechanism for damage detection. An electromagnetic wave, very similar to those which carry radio and television signals through free space, can be propagated from the cable headend down the coaxial "pipe" with relatively little interference from outside signals or to outside receiving equipment. This wave can carry as many as 80 to 90 full television signals, depending on the quality of the cable and the amplifiers used in the system.

This electromagnetic wave does, however, lose power as it moves along the cable. In order to deliver strong and relatively noise-free signals to subscribers, amplifiers must be placed in the transmission lines at separations corresponding roughly to a hundred-fold loss in signal power. Since each amplifier introduces a little distortion and noise, there is a practical limit, depending on the cable bandwidth (number of channels), to the number of amplifiers which can be placed in cascade. This limit has been found to be 20 amplifiers for a 400 to 450 megahertz (MHz) bandwidth, giving a range of about 8 miles with larger (7/8 inch) cables. Special techniques are available to carry signals to more distant distribution hubs. These include high-performance coaxial cable, possibly carrying frequency-modulated signals, microwave links, and optical fiber cables, which will be described later in this chapter.

Microwave transmission is frequently used at the higher levels of a cable system, such as to carry signals from a headend to a fairly distant hub. The popular AML microwave technique, described later in this chapter, is usually operated in the 12.7 to 13.2 gigahertz (GHz) band. Microwave can be economical to set up and can reach more than 20 miles in a single leap. In some geographically dispersed cable systems, the entire headend-to-hub network may be microwave. Microwave is also used to link different systems in "advertising interconnects" (see Chapter 1) and other clustering arrangements.

Optical fiber is still much less prevalent than coaxial cable and microwave. An optical fiber is a very thin solid glass pipe which carries light-wave signals. The refractive index of the glass varies from the core to the outside, making it possible for light waves to propagate down the pipe without leaking out the sides. The light waves can be modulated (varied in some way) with video or other information.

Optical fiber has the important advantages of low loss, so that amplifiers or digital repeaters can be spaced far apart; freedom from outside influences; and resistance to signal theft. Light-wave communication systems have, however, been more expensive than coaxial cable systems, and they are not, for the analog signals and relatively simple hardware used in

(a)

(c)

(b)

Fig. 2.5: Transmission media found in CATV systems. (a) Coaxial cable (Courtesy M/A-COM Company.) (b) Cable-powered AML microwave receiving station. (Courtesy Hughes Microwave Communication Products.) (c) Optical fiber. (Courtesy Times Fiber Cable Company.)

cable systems, capable of carrying as large a number of channels. They are most often used in video feeds for the headend and sometimes for connections to hubs. On the rare occasions when subscriber distribution hubs (Fig. 2.2) are used, fiber may be appropriate for carrying small numbers of channels to subscribers.

Other transmission media, including free-space light-wave communications used within a room or from building to building, are available for appropriate applications. Infrared links are commonly used between wireless control consoles and television sets or other terminals.

Cable is considered a "wide-band" medium because it can accommodate signals which extend over a wide range of frequencies (Fig. 2.6). Unlike a telephone line, with its typical 3000 Hz bandwidth, a coaxial cable can pass a bandwidth starting below 5 MHz and extending, in the newest systems, as far as 550 MHz, although the amplifiers in most existing systems are good only to 400 MHz.

Each CATV channel is 6 MHz wide, and accommodates one full color television signal (see Appendix 1). A television signal is carried with the same amplitude modulation format as is used for over-the-air broadcasting, so that signals brought in for cable transmission need only be "translated" in frequency. Fig. 2.7 illustrates the 54 channel assignments between 54 and 402 MHz. Channels 2–13 have the same frequency assignments as VHF television, but a local television signal is usually translated to a different cable channel to avoid any possibility of "ghosts" on viewers' screens from ingress (leakage into the cable wiring) of the broadcast signal. Most of the gap between channels 6 and 7 is filled by a set of midband channels. Superband and hyperband channels appear above channel 13.

There is plenty of room in a 400 MHz system for local and some "imported" VHF television signals, pay channels, some UHF signals, news and information channels, and specialized programming for small audiences. As we will discuss later in this chapter, the cable operator supplies a "converter" to translate some or all channels to frequencies which can be received by ordinary television sets. Some newer "cable-ready" receivers tune all cable channels.

All in all, a CATV transmission system relies heavily on coaxial cable of various sizes and capabilities, but incorporates other transmission media and

Fig. 2.6: A comparison of the frequency band passed by the standard narrow-band telephone channel with the vastly larger band carried by a standard CATV 400 MHz coaxial cable.

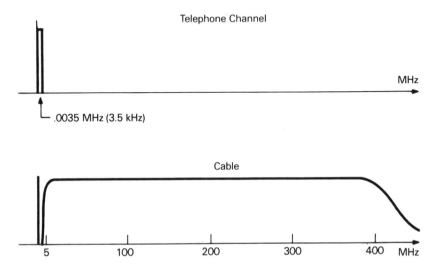

electronic components which are just as important. It may be a good idea to begin thinking of a physical cable system as a more general regional wideband transmission network, with an emphasis on video distribution, using whatever transmission media work best for the several different transmission elements of the network.

Down Through the Cable System: Satellite-Supplied Programming

A cable system begins with programming. Fig. 2.1 shows programming and control signals originating, or being brought in through the cable headend. This programming may include television signals taken "off the air," signals relayed by microwave radio or other means, programming received from communication satellites, material on videotape, and television originating in local studios.

Of all these sources, satellite-supplied programming is by far the most important to present-day CATV systems. The powerful advantage of the satellite medium is that it can be used to distribute programming from a single transmitting earth station to cable systems over a wide area of North America. All of the major program services, which are described in Chapter 3, use satellites.

The satellite link from program supplier to TVRO (television receive-only) station at the cable headend is sketched in Fig. 2.8. The parabolic dish of the receiving antenna is actually a reflector concentrating received signal energy onto the actual antenna, which is a small, horn-shaped device suspended above the antenna or, as seen in Fig. 1.3, underneath a second, smaller reflector. Careful attention is required to frequency coordination with existing terrestrial microwave links (Fig. 2.9) to avoid interference from those links.

Communications satellites are placed in *geostationary* earth orbit 22,300 miles above the equator. Only in this one ring above the equator will the

Fig. 2.7: CATV channels in the frequency spectrum up to 400 MHz. Channels 2 through 13 are the standard VHF channels used in broadcasting. Older cable systems extend only to 300 MHz, while some newer systems reach to 550 MHz. A simpler numerical nomenclature has been proposed.

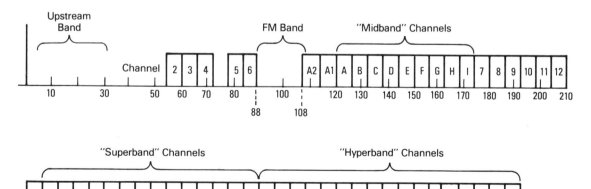

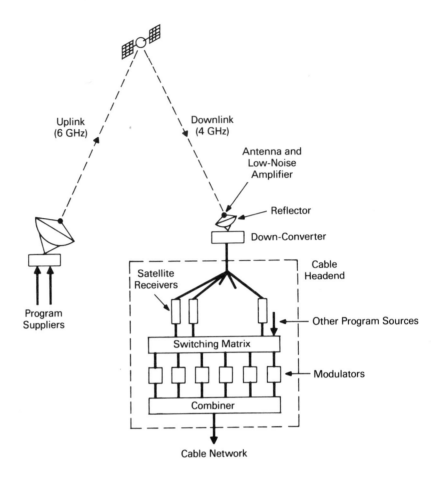

Uplink
(6 GHz)

Downlink
(4 GHz)

Antenna and
Low-Noise
Amplifier

Reflector

Down-Converter

Cable
Headend

Satellite
Receivers

Other Program Sources

Switching Matrix

Modulators

Combiner

Program
Suppliers

Cable Network

Fig. 2.8: *Uplink to a C-band satellite from a distributor of video programming, and downlink to a cable system earth station. The frequencies used in the downlink differ from those used in the uplink to avoid self-interference problems. The receiving "dish," 3 to 5 meters in diameter, is mostly a reflector which concentrates energy on the small antenna (or second reflector) mounted above the dish.*

centrifugal and gravitational forces on a satellite traveling at the earth's rotational speed, and thus hovering stationary above a single point, exactly cancel each other. A stationary satellite can be continuously viewed by a relatively low-cost fixed TVRO antenna. A CATV operator can set up one or a few fixed antennas to view the satellites carrying programming which that operator wants to buy.

The received signal from a satellite is too weak for direct use, so it is strengthened in a low-noise amplifier (LNA), and converted downward in frequency to a band more convenient for transmission to the satellite receivers in the headend. In the headend, one receiver (Fig. 2.10) is provided for each channel to be extracted from the received signal, which may contain as many as 24 separate video channels.

It is difficult to overstate the significance of satellite distribution. For the modest cost, perhaps $15,000 plus $2,000 per channel, of a reasonably good receive-only earth station, a cable operator can access a large number of national distribution networks offering a very wide range of video material.

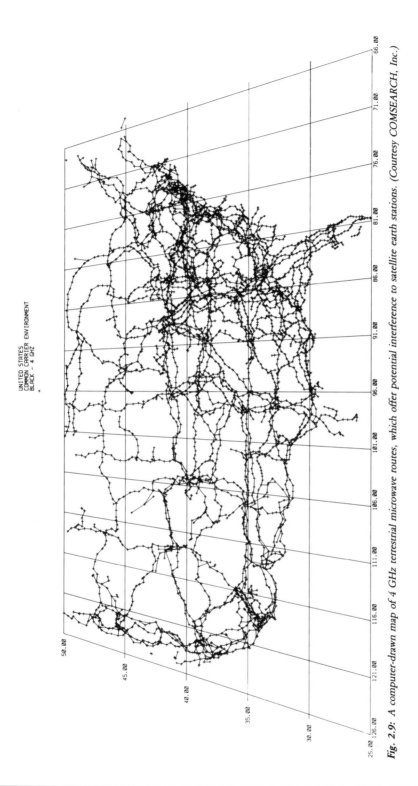

UNITED STATES
COMMON CARRIER ENVIRONMENT
BLACK - 4 GHZ

Fig. 2.9: *A computer-drawn map of 4 GHz terrestrial microwave routes, which offer potential interference to satellite earth stations. (Courtesy COMSEARCH, Inc.)*

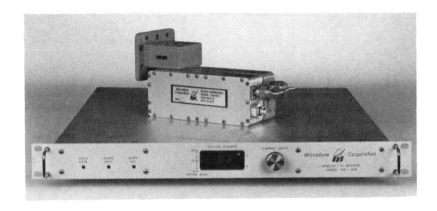

Fig. 2.10: The Microdyne 110BDC/1100DCR receiver, with associated frequency converter. (Courtesy Microdyne Corp.)

From the program distributor's point of view, a satellite video channel, expensive as it may be (about $2 million per year), is a cheap way of achieving instantaneous nationwide distribution. There are concerns about unauthorized reception, or "piracy," of satellite-delivered signals, which are available to at least one million U.S. owners of backyard TVRO stations, although this practice is completely legal. Major program suppliers will soon be scrambling their transmissions, with the annual market for satellite signal decoders forecast at $7.5 billion in the 1990's. Individual earth station owners should be able to purchase scrambled programming, but a nationally coordinated authorization system may be required (see Chapter 10).

A communications satellite, like the now-obsolete RCA Satcom I shown in Fig. 1.2 of Chapter 1, is basically a collection of signal relay stations (transponders) in the sky. Many transponders on board a satellite may share a common pair of receiving and transmitting antennas, but operate on different carrier frequencies. The satellite also carries a field of solar cells to convert sunlight into the electrical energy used by the transponders, and miniature rockets for small adjustments in orbital positioning. Most U.S. commercial communications satellites have been launched by Delta rockets, and some by the European Arianne rocket. The space shuttle, despite the tragic loss, in January 1986, of the Challenger, will carry out many launchings in this decade (Fig. 2.11).

A substantial number of U.S. domestic communication satellites (Fig. 2.12), communicating within the 4–6 GHz *C* frequency band used for both commercial communications and video program distribution, will be in operation in this decade. The gradual reduction of orbital spacing to 2° will allow more satellites to be packed into the available orbital arc, but will require upgrading of less expensive terrestrial stations which cannot "see" sharply enough to clearly separate satellites spaced at 2°.

Table 2.1 describes the 1985 transponder occupancy on several satellites commonly used to supply CATV systems. There is no permanence, however, to these assignments. The combination of a substantial increase in the

(a)

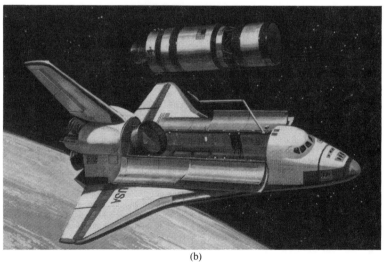

Fig. 2.11: *Delta rocket and space shuttle satellite launch vehicles. The Arianne rocket of the European Space Agency is a third major launch vehicle. The reusable space shuttle carries a satellite, atop a smaller rocket, to a low-altitude orbit, from which it is launched into its permanent high-altitude orbit. [(a), courtesy RCA; (b), courtesy NASA.]*

(b)

Table 2.1: Transponder occupancy on principal U.S. satellites carrying cable programming.*

GALAXY 1	Transponder	Polarization
Nashville Network	2	V
WGN (Chicago)	3	H
Disney (E)	4	V
Showtime (E)	5	H
SIN	6	V
CNN	7	H
CNN Headlines	8	V
ESPN	9	H
The Movie Ch. (E)	10	V
CBN	11	H
Home Team Sports	12	V
C-SPAN	13	H
The Movie Ch. (W)	14	V
WOR (N.Y.)	15	H
WTBS (Atlanta)	18	V
Cinemax (E)	19	H
GalaVision	20	V
HBO (E)	23	H
Disney (W)	24	V

Subcarrier Services	Carried On
Nashville audio	Nashville
Electronic Program Guide	WGN
Moody Broadcasting	WGN
Cable Sportstracker	WGN
WFMT (classical)	WGN
Disney audio	Disney
CNN radio network	CNN
The Movie Channel stereo sound	The Movie Channel
The Greek Network	WOR
The Italian Network	WOR

SATCOM F3	Transponder	Polarization
Nickelodeon (E)	1	V
PTL Satellite	2	H
Financial News Network & Cable Sports Network	4	H
Dodgervision	5	V
SPN	6	H
CBN Cable Network	8	H
USA Cable Network	9	V
Showtime (W)	10	H
MTV	11	V
HBO (W)	13	V
VH-1	15	V
HTN & The Learning Channel	16	H
Lifetime	17	V
EWTN & Reuter Monitor Service & National Jewish TV	18	H
C-SPAN	19	V
BET	20	H
The Weather Channel	21	V
MSN & USA Blackout Network	22	H
Cinemax (W)	23	V
A & E	24	H

*Network feeds and most other non-CATV programming are not shown. Most satellites carry two transponder channels in each of the 12 available frequency slots through use of vertical (V) and horizontal (H) signal polarizations. Many program services have "East" and "West" services, similar except for time shifting. Audio and text channels, some of which are listed here as examples, are transmitted via subcarrier channels squeezed in with video signals, or sometimes (for text) in vertical blanking interval (VBI) channels. (Source: *Satellite Dealer*, June 1985.)

Table 2.1: *Continued.*

Subcarrier Services	Carried On
Star ship adult contemporary	SPN
Star ship country	SPN
Star ship comedy	SPN
Star ship 50s & 60s	SPN
Star ship big bands	SPN
Star ship jazz	CBN
ESPN affiliate information	ESPN
VH-1 stereo sound	MTV
USA cable network stereo	USA

COMSTAR D4	Transponder	Polarization
SelecTV	13	V
Country Music TV	18	H
American Extasy & Nostalgia Network	19	V
Playboy	20	H
KTVT (Dallas)	22	H

WESTAR 5	Transponder	Polarization
University Network	2	V
Pro Am Sports	14	V
Penn National Racing	15	H
FNN	21	H
Meadows Racing	22	V
Fantasy Network	24	V

WESTAR 4	Transponder	Polarization
Word of Faith	2	V
Atlantic City Racing	6	V
Catholic Telecommunications Network	11	H
Music Magazine	19	H

SATCOM F4	Transponder	Polarization
Bravo	2	H
Nickelodeon (W)	4	H
National Christian Network & Prime of Life Network	7	V
SportsVision	9	V
American Movie Classics	10	H
Home Sports	11	V
Playboy	12	H
New England Sports	13	V
BizNet	15	V
Silent Network	16	H
TBN	17	V
WPIX (NY)	19	V
Odyssey	23	V

Subcarrier Services	Carried On
Bravo stereo sound	Bravo
In Touch (reading for blind)	Silent
Star ship progressive rock	Silent
Satellite Jazz Network	TBN

number of satellites and the development of more economical and better performing earth station equipment will keep the satellite distribution business changing for some time.

Off-the-Air Signals and Local Origination

Satellite distribution is preeminent, but the alternative avenues of program supply also have some importance. "Off-the-air" reception of local TV broadcasts supplies many of the channels in the "basic" cable service offered to subscribers, and distant signals may be brought in by microwave. It is unfortunate that off-the-air cable signals are not always of the best quality, and subscribers in major metropolitan areas can sometimes do better with their own antennas.

Studios for local origination (Fig. 2.13) are found in all the larger cable systems and many smaller ones. Public access cable channels and studio facilities to support them are provided for use by community organizations and were once required by the FCC. Cable headends may also use videotape as a program source.

Another kind of programming provided in many cable systems is readable text, such as program guides, stock market quotations, and sports results. Some of these text programs are brought to the headend piggybacked on television signals or by telephone line, while those that are truly local are produced in the cable operator's text entry equipment. In most cases, text programming is not sent to the viewer as teletext (see Appendix 2), which

Fig. 2.12: U.S. C-band communications satellites, in geosynchronous orbit 22,300 miles above the equator, as authorized by the FCC in May 1983. The numbers indicate location in west longitude. The FCC intends to gradually reduce the longitudinal spacing between satellites to 2°, allowing a further increase in the number of satellites.

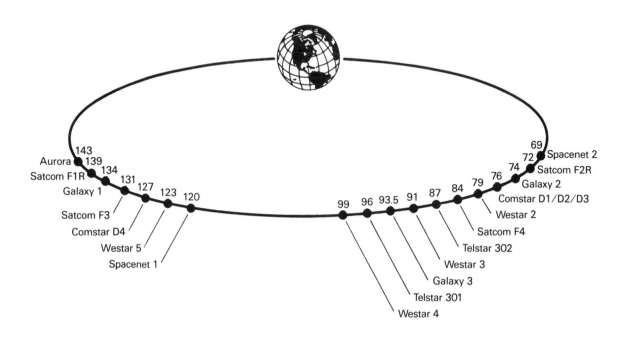

Fig. 2.13: *Children producing programming in a local origination studio. (Courtesy Heritage CableVision.)*

requires an additional decoder at the subscriber location, but by the simple but bandwidth-inefficient method of creating a video signal, in a video character generator, at the headend, and sending it down a spare cable channel. More will be said later in this chapter about text and data transmission to subscribers.

As cable systems become more integrated into the overall communications infrastructure, additional wire and microwave links with communication users and the outside world will develop. Aside from the cost-motivated "clustering" described in Chapter 1, many franchised cable systems are attempting, sometimes for business reasons and sometimes under government prodding, to join in urban and regional networks.

In Columbus, Ohio, for example, the "Industrial Cable Interconnect System" will connect the city's five cable components with city hall. This is to be followed by an "Institutional Cable Interconnect Network" which will allow public institutions, and possibly other users, to communicate with each other. Cable operators are beginning to bid on proposals from government agencies and private companies for dedicated communications facilities. As commercial communications users are increasingly attracted to cable (among other possibilities) for bypass of telephone company facilities or special-purpose communications, the web of interconnections may become more complex and some cable systems will look less and less like broadcasting trees.

Signal Processing in the Headend

Signals from the various distant and local sources are put together in the cable headend. Under the control of operators in a central control room (Fig. 2.14), program sources are assigned to channels within the cable frequency spectrum. This is done by a group of modulators followed by a passive combiner (no amplifiers), as shown in Fig. 2.15. Because of the self-generated harmonic and intermodulation distortions described later, techniques such as the harmonically related carrier (HRC) system, which forces the picture carrier frequencies in the different cable channels to have a strictly proportional and phase-locked relationship, may be used to concentrate some of the distortions at the carrier frequencies, where they are subjectively less disturbing. There is, unfortunately, an incompatibility with some converters and TV sets which cannot detune to the new carrier frequencies, as well as other disadvantages arising from the inflexibility of this solution. An alternative and very similar incrementally related carrier (IRC) system eliminates the detuning problem, but retains the other difficulties.

Steps may also be taken to minimize the potential damage from ingress of strong signals from *outside* the cable system, e.g., local broadcast signals. These signals get into the cables, which are supposed to be completely shielded, through faulty or poorly maintained connectors and wiring. To avoid problems, cable systems may not use their channels at broadcast frequencies for video programming at all, but may instead use them for character-generated text programming, which is less sensitive to the interference. In general, cable operators put their most valuable pay programming in channels which are the least susceptible to harmonic distortions and ingress noise, which is why a subscriber may notice a distinct difference in received signal quality between those programs and a favorite broadcast signal carried in a less desirable cable channel.

Fig. 2.14: A central control room in a cable headend. (Courtesy Warner Amex Cable Communications.)

As Fig. 2.7 suggests, the residential cable transmission bandwidth of 5 to 300, 400, or 550 MHz is not wholly dedicated to transmissions downstream to the subscribers. The bandwidth is divided into one part (50 MHz up) used for downstream transmission and a part (5–30 MHz) dedicated to upstream (return) transmissions, as shown in Fig. 2.16, although not all cable systems have installed the two-way amplifiers and signaling system required to actually realize two-way operation. The institutional cable is more likely to have a mid-split near 112 MHz, or a high split at a still higher frequency.

The downstream bandwidth is sufficient for 35–70 video channels, and the upstream bandwidth, if used for data, can theoretically accommodate a data rate of tens of megabits per second, although it is rarely utilized for more than a small fraction of its capacity. An extra cable can alternatively be provided for upstream communication, obviating the need for bandsplitting equipment, but one-cable bandsplitting is usually adequate to meet the still limited demand. The cable headend may also accommodate computers and other equipment for managing addressable converters (described later) and interactive services.

The Headend–Hub Connection

Fig. 2.15: The cable headend modulator-transmitter function for placing video signals in separate channels and driving the trunk cable(s). The diplex filter separates the upstream from the downstream frequency band.

Cable systems can be built to cover a wide geographic area and/or a large subscriber population by establishing a number of distribution hubs (see Fig. 2.17 on pg. 37). Each hub, which may include the headend, serves a smaller area and a subscriber population of typically ten to fifteen thousand. Signals from the cable headend are carried to the hubs by microwave, coaxial "supertrunk," or optical fiber transmission media.

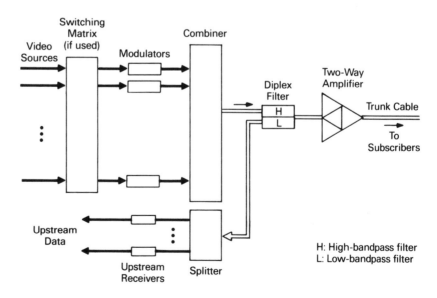

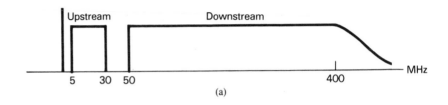

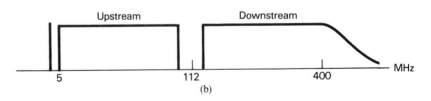

Microwave transmission has the important advantage of avoiding repeat amplifiers, which introduce distortion. A popular technique is the amplitude modulated link (AML). Operating in the 10 GHz band, this system uses the standard television modulation format and makes relatively efficient use of the available frequency spectrum. Low-power AML systems can carry all cable channels in one transmitter signal. Only one receiver needs to be used, and the received signal is simply frequency translated down to the cable band. High-power AML systems, with longer range, may need a separate transmitter for each video channel.

Microwave transmission can suffer from dropout problems in heavy rain, but practical experience has shown that outages are infrequent and brief. Frequency modulation microwave is more robust and can go a greater distance, but requires several times the transmission bandwidth per channel.

Coaxial supertrunk is a larger and higher quality coaxial cable, equipped with special low-distortion amplifiers. It is less expensive to install than microwave, for a substantial number of channels, but cannot deliver as clean a signal over significant distances. Newer "feed-forward" amplifiers, which minimize the intermodulation products (interference from mixing of the different signals on the cable) caused by small amplifier and wiring nonlinearities, are very helpful in trunking. They are also important as amplifiers throughout the newer 550 MHz cable systems, where the increase to 70 channels would otherwise cause a large increase in intermodulation products.

Optical fiber is seen by some industry planners as a very promising trunking medium. Tests reported in 1984 indicate that eight or more frequency-modulated, frequency division multiplexed channels can be carried 20 kilometers (km) without amplifiers on single mode fiber, which uses solid-state lasers for the light sources. The major advantage in signal transmission quality in comparison with coaxial cable is elimination of much of the distortion introduced by amplifiers. The cost of fiber may be

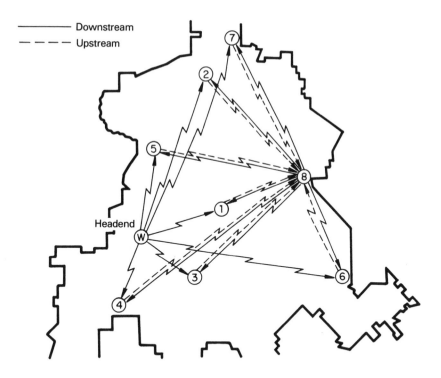

Downstream
------ Upstream

Headend

Fig. 2.17: The nine-hub system, including the headend, of the (former) Warner Amex system in Dallas. Hubs are interconnected by microwave links. (Courtesy Warner Amex Cable Communications.)

just becoming competitive with AML microwave and coaxial supertrunk cable, especially where long runs and very high signal quality are important, although analog transmission on fiber has its own distortions.

Digital optical transmission systems, widely used in the telephone network, may not yet be economically practical for CATV applications, but sooner or later will become the preferred candidates for supertrunking within and between cable systems. Digital regenerators, required at large intervals, will maintain signal integrity over almost unlimited distances. At rates of the order of 90 Mbits/s per video signal, these digital systems will provide transmission of extremely high and unvarying quality.

Cable Distribution from the Hubs

Three or four trunk cables branch out from each hub, usually carrying the same programming. In systems set up for two-way communication, with the frequency spectrum split into the 5–30 MHz upstream and 50 MHz up downstream bands described earlier, the trunk amplifiers (Fig. 2.18) will have separate ''upstream'' and ''downstream'' amplifier sections.

Upstream signaling is difficult for a number of reasons. One is noise ingress (pickup) due to faulty connectors and other imperfections. The tree-like architecture of the cable system tends to combine the collected disturbances from different branches into a high noise level, although it may

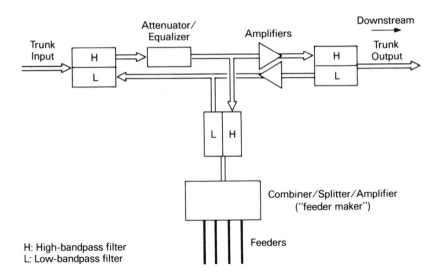

Trunk Input

Attenuator/ Equalizer

Amplifiers

Downstream

Trunk Output

H
L

H
L

L | H

Combiner/Splitter/Amplifier
("feeder maker")

H: High-bandpass filter
L: Low-bandpass filter

Feeders

Fig. 2.18: An intermediate trunk bridger/amplifier station supplying four feeder cables from a trunk cable. Downstream signals are "split" into four identical copies passed to each feeder, and upstream signals from the feeders are combined into one upstream signal for the trunk cable. Filters prevent downstream signals from going upstream and upstream signals from going downstream.

not be intolerable in properly designed and maintained systems. Switched bridges, described later, can be used to isolate noisy or faulty parts of the cable system. A second problem is that of keeping the total power of the upstream signal within a tolerable range when signals in different upstream channels may or may not be "on." This problem discourages the development of very wide-band upstream transmission.

Alternating current (ac) power for the trunk amplifiers and other line units is inserted into the cable at various points and carried on the cable along with the information signals. It does not interfere with the information signals because its frequency (60 Hz) lies far outside of their transmission bands. Not all cable systems have battery standby power, and those that do not may go "down" when there is a power outage. However, many major systems have full battery backup, and this is becoming a more common practice throughout the industry.

Trunk cables are further branched into feeder cables at intermediate and terminating *bridgers* (Fig. 2.1). As illustrated in Fig. 2.18, the intermediate bridger typically drives four feeders, and can be built to accommodate both downstream and upstream transmission. Fig. 2.19 pictures a typical two-way bridging station. It is possible for bridging stations to be addressable from the cable headend, such that the upstream portions of the feeders are switched in or out on the command of the headend, with a packet of data carrying the message to the bridge. An addressable bridger is illustrated in Fig. 2.20. This makes it possible for the headend to listen to a relatively small part of the cable system, typically covering 100–200 subscribers. Originally designed to facilitate isolation of parts of the cable network in order to locate faults, addressable bridgers have been used in some interactive systems to reduce upstream data transmission problems, as we will discuss later on.

Fig. 2.19: A typical trunk
bridging station. (Courtesy
Scientific Atlanta Company.)

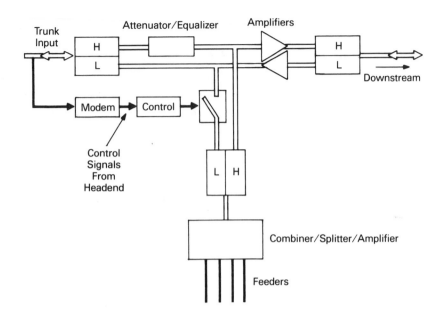

Trunk Input

Attenuator/Equalizer

Amplifiers

H
L

H
L

Downstream

Modem

Control

Control Signals From Headend

L H

Combiner/Splitter/Amplifier

Feeders

Fig. 2.20: A trunk bridging station with addressable upstream ports. The upstream channels of the feeder cables are switched in or out under the control of data sent from the cable headend.

The feeder cables are "tapped" near each subscriber location. A directional coupler in the tap (Fig. 2.21) provides a subscriber, through his drop cable, with a small part of the signal power in the feeder cable, but enough to ensure strong and clear reception. The tap incorporates a directional coupler which strongly attenuates echoes and transmissions from subscribers and keeps them from interfering with other subscribers' reception of the downstream signal.

The proper attenuation of the signal strengths from feeder cable to subscriber drop cable varies, depending on the signal strength at that point in the feeder cable, which in turn depends on the output level of the last amplifier and the loss elements (cable, splitter, and previous taps) incurred after that. The object is to provide each subscriber with the same signal strength of about 12 dBmV (decibels above a millivolt, or 4 mV). Taps are manufactured with a range of tap values (attenuations), typically from 3 to 40 dB. Engineers prefer the logarithmic decibel scale to ordinary power attenuation factors, which in this case would be 2 to 10,000. The undesirable but inevitable insertion loss which the tap places in the feeder cable is inversely related to tap value, and typically ranges from 0.2 to 4 dB.

Converters

Once inside the subscriber's residence, the drop cable either connects directly to the subscriber's television set or goes through a *converter* and/or other subscriber electronics (see Fig. 2.22 on pg. 42). The converter (Figs. 2.23 and 2.24) is necessary if the cable system supplies more than channels

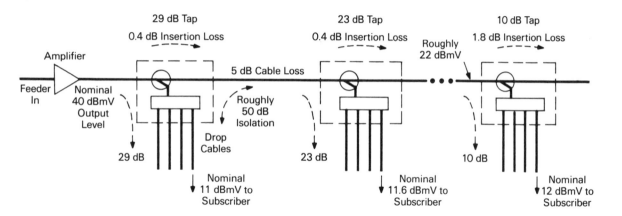

Amplifier

Feeder In

Nominal 40 dBmV Output Level

29 dB Tap
0.4 dB Insertion Loss

5 dB Cable Loss

Roughly 50 dB Isolation

Drop Cables

29 dB

Nominal 11 dBmV to Subscriber

23 dB Tap
0.4 dB Insertion Loss

Roughly 22 dBmV

23 dB

Nominal 11.6 dBmV to Subscriber

10 dB Tap
1.8 dB Insertion Loss

10 dB

Nominal 12 dBmV to Subscriber

Fig. 2.21: Taps of differing value along a feeder cable. The downstream transmissions are attenuated by whatever values necessary to obtain a proper subscriber signal voltage level of about 12 dBmV. An attenuation of 10 dB corresponds to 10-times power reduction, 20 dB to 100-times, and 30 dB to 1000-times. Return transmissions from a subscriber are similarly attenuated on their way upstream, but are reduced by a large additional factor, perhaps 50 dB (100,000-times), upon entering the drop cables of other subscribers, and thus cause negligible interference.

2–13. Only "cable-ready" television sets can tune the higher frequency cable channels. If additional channels, such as pay channels, are supplied by the cable operator, they must be translated in frequency down to the frequencies employed by one or more of channels 2–13 (or sometimes up to UHF frequencies). A *block converter* [Fig. 2.23(a)] moves a contiguous block of channels, usually from the midband group (Fig. 2.7), down to the frequencies of channels 7–13 (for seven channels) or up to UHF for all channels. With the converter off the subscriber receives the normal channels; with the converter on, these signals are replaced by the translated cable channels.

A *general converter* (Figs. 2.23(b), 2.24) makes a one-at-a-time shift of a cable channel down to a single VHF channel, sometimes via an intermediate unmodulated baseband or UHF stage, and can be designed to convert as many channels as desired. A baseband intermediate stage is convenient for descrambling or stripping teletext from a television signal, but a UHF intermediate stage (Fig. 2.25) can reduce the level of intermodulation distortion resulting from imperfections of the semiconductors in the converter.

As suggested earlier, a "cable-ready" television set has a built-in general converter which can directly receive all cable channels up to some frequency limit. These sets have, unfortunately, been incompatible with scrambled pay channels, as discussed in the next section.

Addressable converters (see Figs. 2.26, 2.27 on pp. 45 and 46, resp.) are converters that can make channels available or unavailable in accordance with instructions sent in data packets from the cable headend. They are part of the sales system, in addition to carrying out the normal frequency translation function of a converter. As suggested in Fig. 2.27, a frequency translator, controlled by the subscriber, picks out a specific channel. If the program is in one of the pay categories, it is scrambled (as described in the next section) and can be unscrambled only under control of data sent from the headend. Each subscriber has a unique electronic address, so that the cable headend has the

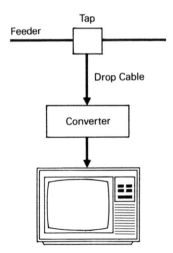

Fig. 2.22: *A converter provided to the subscriber enables reception of a large number of video channels. The converter can be combined with electronics for interactive communications.*

Fig. 2.23: *Two types of converters are shown: (a) The block converter, which translates a contiguous block of channels to a range tunable by the receiver. (b) The general converter, which selectively translates a cable channel to one VHF channel not used for local broadcasting, e.g., channel 3 in the New York metropolitan area. The baseband converter illustrated here converts the selected channel down to an unmodulated composite video signal before remodulating it to a channel which can be received by the TV set.*

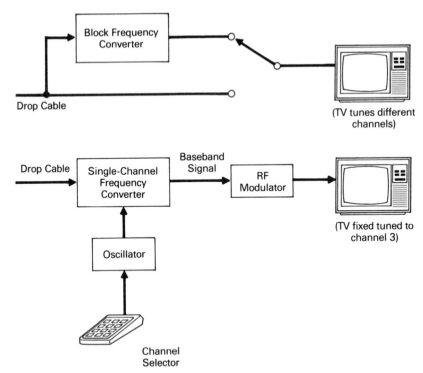

Fig. 2.24: *A 36-channel tunable converter. (Courtesy General Instrument Corporation.)*

Fig. 2.25: *Design of a general converter with UHF intermediate stage (from [7])*

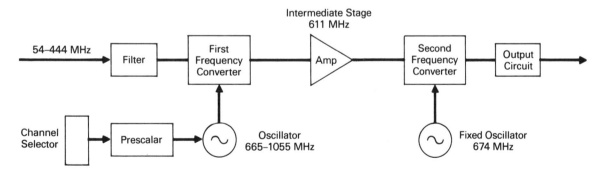

CABLE-READY

opportunity to provide viewing authorizations to each addressable converter.

This system makes it easy and inexpensive for the cable operator to change the mix of pay services delivered to a customer; it is not necessary to visit the subscriber's home. "Pay-per-view" delivery of individual programs becomes relatively easy, although the subscriber has to request the program by mail or telephone unless the cable system is interactive. The cost of an addressable converter is greater than that of a nonaddressable converter, but the difference may be only about $20.

Off-premises addressable converters (Fig. 2.28) are also available, with one or several channel selection modules in the subscriber's home. This extremely miniature subscriber switching "star" is intended to prevent signal theft. Scrambling of pay signals is unnecessary, or so it is believed, because the signals are available only on the feeder cables. Some of the interactive systems described later include off-premises converters as an extra feature.

Fig. 2.26: An addressable converter system includes the converters installed at subscriber locations, and a control installation where subscriber authorizations are entered or altered. (Courtesy General Instrument Corporation.)

Cable System Security: Supplying Programming Only to Those Who Pay

Three techniques for restricting pay channel reception to subscribers who are paying for them are commonly used in cable systems. They are *jamming*, *trapping*, and *scrambling*, all explained below. The object is to present a severely distorted picture, or no picture at all, to a subscriber tuning to a pay channel (or one of a packaged tier of pay channels) who is not authorized by the cable operator to receive that programming. The denial of programming should also, of course, apply to a non-subscribing tapper of the cable network.

The *jamming* system inserts an interfering noise, at the cable headend, into a supposedly non-critical region within the frequency band occupied by the pay signal. The carrier frequency of the interfering noise is usually 2.25 or 2.5 MHz above the carrier frequency of the television picture, where picture sideband information is transmitted. A notch filter [Fig. 2.29(a)] supplied by the cable operator removes the interference before the signal enters the converter. Unfortunately, it also removes a small part of the desired video information, reducing the contrast slightly. This "soft picture" is usually not considered objectionable. Jamming is not a strong security system, because bootleg notch filters can be easily built and installed, and it is not used in newer cable systems.

The *trapping* system [Fig. 2.29(b)] uses a filter which is installed at the beginning of the subscriber's cable drop outside his residence, preferably high up on a telephone pole. This filter suppresses the entire pay channel, and is supplied to all subscribers *except* those taking the pay channel. It is more secure than the jamming system and is widely used, although it is still susceptible to fraud through simple removal. In urban apartment buildings,

Fig. 2.27: Structure of an addressable converter, giving the cable operator immediate remote control of the availability of different channels to a subscriber. Scrambled transmissions (and a descrambler in the converter) are employed to prevent signal theft through use of ordinary converters.

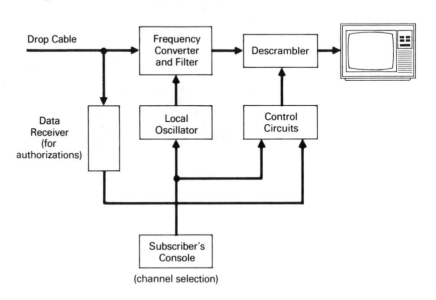

Fig. 2.28: An off-premises addressable system. The addressable switching units for several subscribers are housed in an outdoor enclosure; single units may be suspended from the cable strand. (Courtesy Blonder-Tongue Laboratories, Inc.)

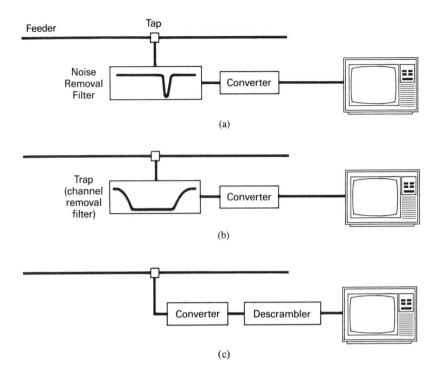

(a)

(b)

(c)

Fig. 2.29: Three alternative systems for denying pay services to subscribers who have not paid for them. (a) Jamming: narrow-band noise inserted at cable headend into pay channel is removed by notch filter at subscriber's location. (b) Trapping: entire pay channel is removed before entering non-subscriber's drop cable. (c) Scrambling: authorizations in the converter allow or disallow descrambling of a pay channel.

the traps are sometimes placed in easily reachable locations inside the buildings.

Traps may, for special purposes, be directly accessible to subscribers. Fig. 2.30(a) shows a disposable trap sold for one-time use to receive a pay-per-view event, and Fig. 2.30(b) illustrates a key-lock trap for "parental control" purposes. Fig. 2.30(c) shows, for comparison, the usual outdoor trap.

The *scrambling* or *coding* system [Fig. 2.29(c)] acts on pay signals or upper-tier channels which are scrambled at the cable headend. The most common scrambling scheme is pulsed sync suppression, which effectively removes the horizontal sync pulses (Appendix 1) and thereby prevents the receiver from locking in on the video waveform. The descrambling operation requires a timing signal, ordinarily sent as part of the sound component of the pay TV signal. Descrambling may be done either on a passband (modulated) signal, or on a baseband composite video signal.

A second scrambling system, also widely used, reverses the luminance component of the signal (video inversion) so that black becomes white and white black. This scheme requires that the signal be in baseband format. The two schemes are often used together as the suppressed sync and video inversion (SSAVI) technique, which has the additional advantage of reducing the signal load on the distribution network and thereby reducing the intermodulation distortion. More advanced schemes, not yet commer-

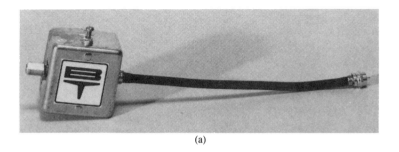

(a)

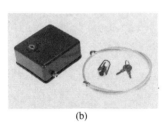

(b)

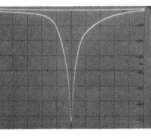

(c)

Fig. 2.30: *Three different trap filters. (a) A disposable trap sold for a single pay-per-view event, with a short-life battery activated by a pull tab just before a viewing event. It connects directly to the antenna terminals of the TV set. (Courtesy Blonder-Tongue Laboratories, Inc.) (b) A key-lock "parental control" trap. (Courtesy Intercept, Inc.) (c) A standard outdoor line trap and its frequency attenuation characteristic. (Courtesy Microwave Filter Company.)*

cially available in residential subscriber equipment, shift, reverse, or exchange picture trace lines. They require substantial electronic memory and computational capability in the descrambler, and will probably not be used until television receivers are fully digitized.

Authorization for pay channels is programmed into the converter, which then will pass to the descrambler only those channels which are authorized. It has already been explained how *addressable* converters can be instantaneously reauthorized from the headend. The converter and descrambler are built together into one secure box. This is a reasonably secure system and is preferred in newer systems, but it is, unfortunately, incompatible with cable-ready television sets which incorporate converters but not decoders. A joint EIA (Electronic Industries Association)/NCTA (National Cable Television Association) committee has developed a decoder interface which would allow use of the tuner and other consumer-oriented features of a television set with a decoder supplied by the cable operator. Newer component television systems, with the television tuner separated from the video monitor, may also allow interposition of an operator-supplied device, or, alternatively, the monitor could be attached directly to a descrambling converter.

The greatest security can be achieved by placing addressable converters outside of the subscriber's reach, as in the subscriber star or off-premises converter described earlier. As was already noted, it is not necessary to scramble the pay channels if the cable carrying all programming is effectively inaccessible to the subscriber.

Text and Data Broadcasting on Cable

In addition to video programming, and sometimes associated with it, many cable systems provide broadcast text and data programming such as captioning, financial news reports, teletext information services and microcomputer software. Digital music may be a future offering, and is already possible with Toshiba "DCAT" equipment. These examples of data communications are explained as services in Chapter 4. The techniques for delivery vary greatly depending on the relative costs of transmission capacity and subscriber receiving equipment.

There are four commonly used approaches, as shown in Fig. 2.31. The first is conversion of text to a television signal in the headend. This is accomplished by connecting a local character generator, or a text stream arriving at the headend from somewhere else, to NTSC television signal generating equipment. The result is a television signal which does not require any additional subscriber equipment beyond that provided for video reception. However, it is a very wasteful user of transmission capacity, since a full television channel is consumed for an information stream which could easily be transmitted in a channel a thousand times narrower. Even in a newer system with 35 or more channels, only a very few channels, such as a program guide and sports results, are likely to be dedicated to this kind of transmission.

The second technique, described further in Appendix 2, piggybacks a data signal on a regular television signal by sending data, instead of luminance information, during the vertical blanking interval (VBI) between transmission of television display fields. If certain encoding and presentation standards are followed, this technique is called teletext. Although a high data rate of 5.72 MHz, chosen for technical reasons described in Appendix 2, is typically used during this interval, the interval is a short one corresponding to the time to trace about 21 horizontal sweep lines during each 525-line television frame. The effective data rate, after further reductions because part of the VBI is dedicated to other uses, is about 200 kbits/second, which is adequate for many services and not to be scorned since it is "free" in the sense of not requiring any additional transmission capacity. It does, however, require special decoding equipment at the subscriber's location, equipment that is still fairly expensive but may become a low-cost built-in option on future television sets. A teletext "magazine" of 200 frames of text and graphics can be delivered in about 10 seconds.

The third data broadcasting technique is full frame teletext in which an entire video channel is used. The same video modulation technique, data formatting standard, and 5.72 MHz/second burst data rate are used as for VBI teletext, but because there is no picture transmission, the 5.72 MHz/s transmission is on continuously rather than intermittently. Thirty times as much data can be sent each second. However, fairly sophisticated receiving equipment may be needed.

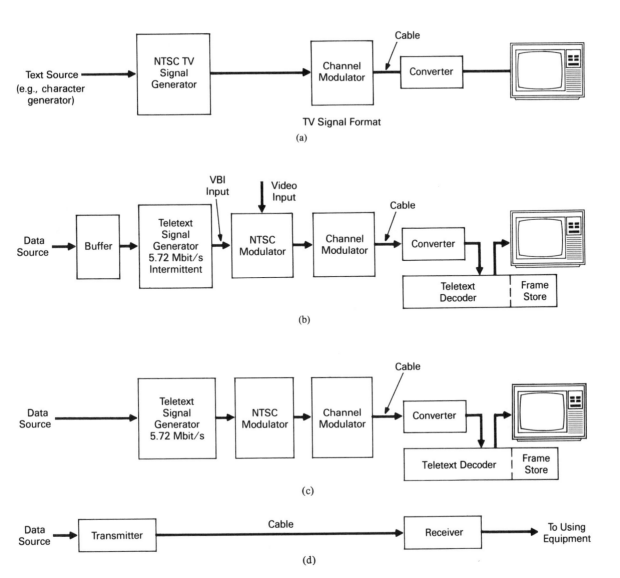

Fig. 2.31: *Four approaches to text and data broadcasting in cable systems. Teletext, linked to television formats, is described in Appendix 2. (a) Conversion to a TV signal. (b) VBI teletext. (c) Full-frame teletext. (d) Independent data broadcasting.*

The fourth technique is direct data transmission in any available downstream frequency band, independent of television or teletext formats. This gives added degrees of freedom which can improve performance, but requires special subscriber equipment, possibly in addition to teletext receiving equipment. The Nabu company offered a system to U.S. cable operators in 1984 which transmitted computer software to paying subscribers at 6.312 Mbits/s over a full video channel. A modulation format called offset quadrature phase-shift keying (OQPSK) was used because of its resistance to noise and distortion, and specially equipped receivers and microcomputers were provided to subscribers.

Interactive Systems and the Metropolitan Area Network

Cable systems have been and remain largely broadcasting systems, distributing programming and occasional control signals (such as instructions to addressable converters) downstream to subscribers. However, many services, including some strictly related to selection of video entertainment as well as the general digital communications services implicit in the concept of a metropolitan area network, require communication from the subscriber to the cable headend or other processing and routing points. For competitive business reasons and advantages of control and convenience, cable operators would like to carry this upstream traffic through the cable system, although hybrid CATV-telephone line systems, described in the next section, have their own advantages. Some examples of services which can be supported by an interactive capability are pay-per-view entertainment, building security, information retrieval, and bulk digital transmission.

In colloquial usage, the term "interactive system" usually refers to a cable system with relatively low-rate data communications between subscribers (or automatic terminals at their locations) and a computer in the headend. The concept may take in some data communication with outside parties reached through the headend, as illustrated in Fig. 2.32. Only a very small fraction of cable subscribers were on interactive systems when this was written. General purpose and higher rate data communications, as between two business offices connected to the cable, is somewhat different in concept and execution from existing interactive systems, and is described in the next section and further in Chapter 4. Bulk transmission of digitized voice is an important application for digital transmission at the T1 rate (1.544 Mbits/s) and higher.

The interactive communication link between a headend and a residential subscriber in the mid-1980's interactive systems is not expected to carry very much traffic. In the QUBE systems of the Warner Amex Cable Company, the first commercial realization of interactive cable, the subscriber is given a small console (Fig. 2.33) with a limited set of pushbuttons. By pressing a certain combination of buttons, a short message is created which, in response to a request sent downstream from a computer in the headend, is sent back *upstream* to that computer. A typical use is for channel selection, but the system has frequently been used for pay per view and for audience polls. For audience polls, the QUBE computer is set up to immediately tally the responses and display the results in the video program which the audience is viewing (see Chapter 4).

Even for this limited interaction, there is considerable technical difficulty in implementing the necessary upstream (and downstream) data communications. The cable operator must install bidirectional amplifiers and bridges, appropriate computer control facilities at the headend, and subscriber electronics designed to communicate with the computer and accept instructions from the subscriber-operated keyboard.

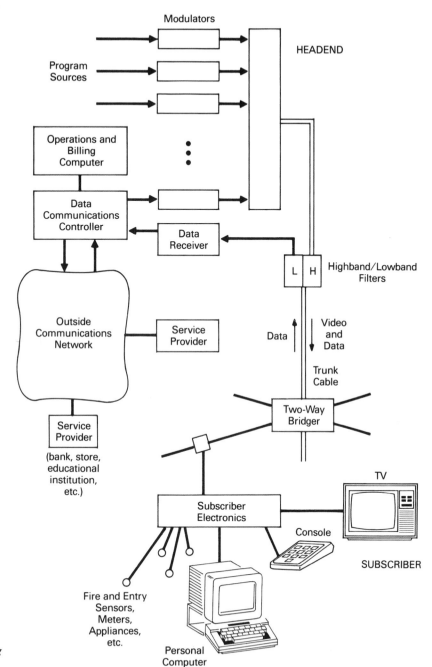

Modulators

Program Sources

HEADEND

Operations and Billing Computer

Data Communications Controller

Data Receiver

Highband/Lowband Filters

L H

Outside Communications Network

Service Provider

Data

Video and Data

Trunk Cable

Service Provider

(bank, store, educational institution, etc.)

Two-Way Bridger

TV

Subscriber Electronics

Console

SUBSCRIBER

Fire and Entry Sensors, Meters, Appliances, etc.

Personal Computer

Fig. 2.32: *Architecture of an interactive cable system, indicating some of the possible services.*

Fig. 2.33: *A subscriber keyboard for the interactive cable system marketed by Pioneer. This polling-type answer back technology was first offered to the public on the Warner system in Pittsburgh in 1977. (Courtesy Pioneer Communications of America, Inc.)*

Most interactive systems use shared upstream and downstream data channels. For the residential cable, the upstream channels lie within the 5–30 MHz band, and the downstream channels within the 120–130 MHz band. Addressable bridgers, if installed in the cable system, shut out the collected noise and interference from inactive parts of the system. Only those subscribers connected through a feeder cable from the single "on" bridge are allowed to communicate with the cable headend at a given time. Switching large sections of the subscriber population in and out is good for noise reduction, but makes it difficult, in contrast with the highly interconnected telephone network, to provide simultaneous service to a substantial number of subscribers. The situation is not quite this bad because each hub of a large system can be operated as an autonomous communications center linked to the headend by whatever extra communications capacity is required.

Although interactive systems are most frequently intended for communication between a subscriber terminal and the headend or an outside party reached through the headend, subscriber to subscriber communication is also possible through the headend. In this case, the upstream data arriving at the headend is simply retransmitted in a downstream data channel. If hubs are equipped with retransmitters and are interconnected, the bottleneck at the headend can be avoided by routing some of the traffic back downstream at the hub level.

There are several approaches to using the common data channels. The one implemented in the QUBE system and others like it is a regular polling arrangement, in which subscriber terminals are sequentially interrogated by the control computer in the headend. Every 5 to 10 seconds, each subscriber terminal is "asked" for a status report and any subscriber-entered information. Working terminals respond with an "I'm O.K." acknowledgment and possibly with a return message. Terminals not responding at all have their addresses printed out for possible service referrals. The return message, if any, may be a programming choice just made by the subscriber, some other pushbutton response from the subscriber, or the report of an automatic device such as a burglar alarm.

The QUBE system uses a data rate of 256 kbits/s in each hub-centered polling group of 10,000–15,000 subscribers, using frequency-shift keyed (FSK) transmissions at a downstream carrier frequency of 124 MHz and an upstream carrier frequency of 25 MHz. The polling cycle takes 10–15 seconds. Although the polling strategy can give more attention to an active subscriber than to an inactive one, it is difficult in a system of this kind to provide enough capacity to a significant population of users for information retrieval, shopping, and other interesting services. The system can, however, be quite effective for low rate program selection and environmental monitoring applications, as described further in Chapter 4.

Polling systems are susceptible to improvement, but in general have not delivered adequate solutions to the two main problems posed by CATV system architectures: how to provide adequate interactive data communications capacity, on demand, to many subscribers simultaneously, and how to minimize the damage from distortion, outages, and noise collection in a network with a tree and branch configuration and too many amplifiers and connections. The proposed answers take many forms, but a common element is frequent *pulse regeneration* in the upstream communication path (Fig. 2.34) to defeat the effects of noise collection.

Pulse regeneration in a digital communication system simply means replacing a distorted or noisy signal pulse with a clean one. Because signal pulses in a digital communication system are created with only a small number of discrete levels, e.g., ± 1 in binary bipolar signaling, and assuming an input pulse is not so distorted that an incorrect decision on its level is likely, the regenerator can clean up the distortion and effectively solve the noise collection problem. Regenerators may be placed at the headend, hubs, bridging points, taps, or wherever branches come together. Packet-switched operation, in which the data stream from each user is broken into small data packets, is often used to increase the efficiency of channel utilization and to make retransmission of damaged data easier.

The answers to the multiple access problem of supplying adequate transmission capacity on demand are different demand assignment schemes which are matched to different kinds of traffic. For steady transmission at relatively high rates, the dedicated frequency channels of frequency division multiplexing (FDM) or dedicated time slots of time division multiplexing (TDM) may be used. For the more "bursty" traffic typical of terminal communications, the choice is more likely to be a random access, or contention system (Fig. 2.35).

In the commonly used carrier sense multiple access/collision detection (CSMA/CD) scheme, a terminal goes "live" with data packets on a common upstream data channel when it has something to send. All packets received by the headend, whether addressed to the headend or to another subscriber, are retransmitted in the downstream data channel. Subscriber

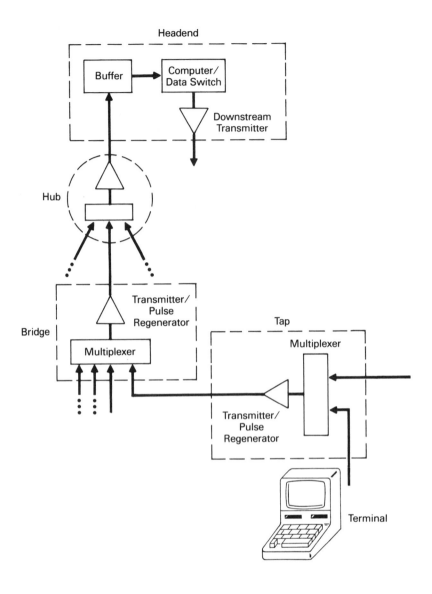

Headend

Buffer

Computer/
Data Switch

Downstream
Transmitter

Hub

Bridge

Transmitter/
Pulse
Regenerator

Multiplexer

Tap

Multiplexer

Transmitter/
Pulse
Regenerator

Terminal

Fig. 2.34: Pulse regeneration as a response to the noise collection problem for upstream data transmission. This can be a time-slot allocated time division multiplexed (TDM) system, or some form of packet-switched system. The downstream facilities are not shown.

stations listen to the downstream data channel to check that other terminals are not transmitting before beginning their own transmissions. Despite this precaution, occasional collisions of packets from different originators are expected and are taken care of by retransmissions. A packet is sensed by all subscribers, but taken in only by the subscriber to whom it is addressed.

One problem with a CSMA protocol on a large cable system is that the probability of a collision is high with a large distance, and thus a large signal propagation time, between stations. It has been suggested that a CATV system be broken in smaller systems, each operating efficiently with its own

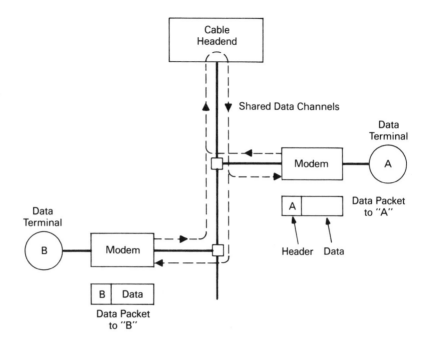

Cable
Headend

Shared Data Channels

Data
Terminal

Modem

A

Data Packet
to "A"

A

Header   Data

Data
Terminal

B

Modem

Data Packet
to "B"

B   Data

Fig. 2.35: *A random access system in which sections of subscriber messages, organized into addressed data packets, are applied to a common upstream transmission channel more or less as they are created. All packets are simply retransmitted by the headend in the downstream channel, in a different frequency band.*

local headend under the CSMA protocol, and each linked to the cable headend by a private connection from its local headend. A network modified in this way could handle both voice and data on a contention basis.

Several commercial systems operating more or less according to these ideas have been introduced in recent years. In Cox Cable's INDAX system, which suspended operation in 1984, messages applied to the cable randomly contended for channel access. Subscriber stations listened to the downstream data channel not only to see if it was clear of sense collisions, but also to receive their own transmissions and check them against the stored originals for errors. If there were errors, the transmission was repeated.

The INDAX systems used up to 40 separate FSK channels upstream and up to 80 downstream, which facilitated adaptive allocation of capacity to subscribers. Each channel operated at the relatively slow data rate of 28 kbits/s, although, if temporarily assigned to one subscriber, this is a very high rate for services such as videotex. The subscribers' data transmitters were frequency agile, allowing switching to alternative channels, but this capability was used only for noise avoidance or long-term load balancing.

The system used a considerable bandwidth: 12 MHz (17–29 MHz) for the 40 upstream channels, and 24 MHz (108–132 MHz) for the 80 downstream channels. The relatively low power used for the downstream data transmissions allowed use of frequencies which would not be available to higher powered signals which could possibly radiate from bad cable connections at high enough levels to interfere with air navigation signals. Upstream ingress noise collection was diagnosed (as in non-INDAX cable systems) by using addressable switched bridgers to isolate troublesome sectors of the cable network. This did not, by itself, correct the problem, and there was no pulse regeneration in the system except at the headend.

INDAX was actually a concept for integrating normal video, one-way teletext, two-way videotex, and "gateway" videotex (to outside services) in one coordinated system. For data services, such as banking or shopping transactions by videotex, assumptions on subscriber usage were made which had a large part in determining the design of the system. As in designs of commercial data communication systems, enough capacity was provided to allow for simultaneous "busy hour" usage by many subscribers with minimal delays. Extensive provision was also made for the security of transactional sessions, and for communication with service providers outside the cable network via data transport networks.

MetroNet, an interactive data communication system intended more as a regional data network bypassing the local telephone company than as an enhancement of CATV, was developed by General Instrument Company and Sytek, Inc. in the mid 1980's. It was described as a metropolitan area network designed to transform a two-way CATV cable plant into a fully-connected local packet-switched network, allegedly inexpensive to implement, and using multiple CSMA/CD channels operating at 128 kbits/s. Each channel could accommodate up to 300 simultaneous users. Twenty data channels, each occupying a 300 kHz bandwidth, were multiplexed within a 6 MHz CATV channel. The system used standardized interface units, a structure and communications protocols very similar to those of commercial local area networks, and digital packet repeaters which both regenerated pulses and retransmitted them downstream.

The digital packet repeaters were located at hubs of the MetroNet system, which could be at the CATV headend, at CATV hubs, or at any branching points in the CATV network. With repeaters adequately deployed, the MetroNet design could solve the noise collection problem.

It was claimed that up to 100,000 subscriber terminals could be supported by a single MetroNet hub, and that the instantaneous data rate of 128 kbits/s could be dedicated to downloading a display frame to a single subscriber, providing the very fast browsing capability which is so important to the future success of videotex services.

Other new systems place even more emphasis on multiple services and on digital pulse regeneration, although they may not have all the capabilities of a regional area network. Omnitel, a design offered by GEC-Jerrold, grew out of the integrated services experiment called Ida which was conducted in Manitoba, Canada in the 1970's. Its services include, and even emphasize, subscriber voice, and its architecture (Fig. 2.36) resembles that of digital subscriber carrier systems used by telephone companies. In this case, the time division multiplexed voice channels are frequency-division multiplexed

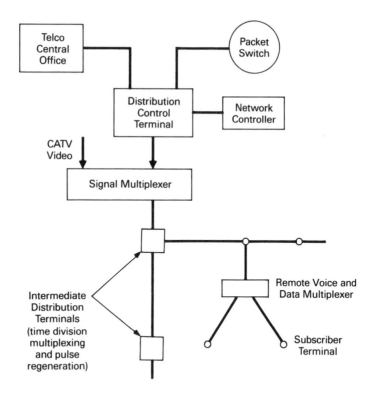

Fig. 2.36: The Omnitel
interactive data communication
system, adding voice and data
services to a cable network (from
Stahlman [6]).

side by side with standard CATV services. Intermediate distribution
terminals, located largely at CATV bridging points, control TDM assignments and carry out pulse regeneration.

Another design for conversion of a CATV system into a full-fledged store-
and-forward packet switching network has been proposed by PacketCable.
In this approach, illustrated conceptually in Fig. 2.37 but not necessarily
identical to the proprietary node architecture of the developers, the
subscribers on a given feeder cable transmit addressed data packets on a
random access basis, e.g., CSMA, to what are effectively miniature
subscriber hubs. Subscriber-generated packets are queued, in their arrival
order, at these entry nodes and at others up and down the cable network,
and the packets, with regenerated data pulses, are transmitted on outgoing
cables.

There are no switches or time slot assignments at the nodes of packet-
switched networks of this kind, only traffic-flow controllers, buffers, and
transmitters which regenerate pulses to maintain signal quality. Packets may
or may not go through the headend; node-to-node crossover links, at major
hub level or any other node level, could be implemented without disturbing
the architecture. Routing strategies can be built into the system to control
data flows and minimize network transit times.

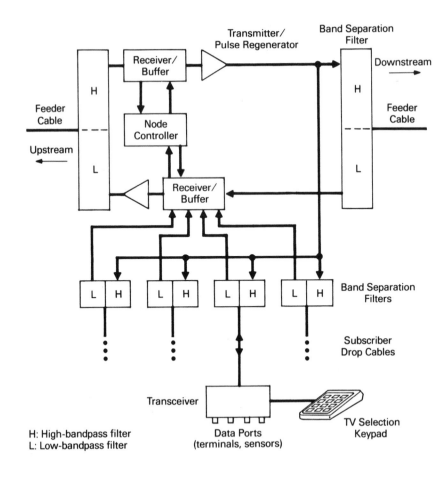

Transmitter/
Pulse Regenerator

Band Separation
Filter

Receiver/
Buffer

Feeder
Cable

Downstream

H

H

Node
Controller

Feeder
Cable

Upstream

L

L

Receiver/
Buffer

Band Separation
Filters

L H   L H   L H   L H

Subscriber
Drop Cables

Transceiver

TV Selection
Keypad

H: High-bandpass filter
L: Low-bandpass filter

Data Ports
(terminals, sensors)

Fig. 2.37: Node at a multiple subscriber tap of a packet-switched data network realized within an urban cable system. The downstream video function is not illustrated, but demand-assignable off-premises converters can be integrated with the data communications functions in this intelligent tap.

Some other proposals for two-way cable do not rely on either bridge switching or on upstream pulse regeneration. Zenith Electronic Corporation's ZVIEW depends on detection of errors in noise-damaged messages and on a robust phase modulation system. It also uses a combination of polling and a random access system known as slotted Aloha. In slotted Aloha, time is broken into slots defined by a master clock. Terminals do not transmit at arbitrary times (after listening for a clear channel), as in CSMA/CD, but rather transmit in the next time slot, without listening for a clear channel. If there is a collision between transmissions from different terminals, errors occur and are detected by the recipients, who request retransmissions. There are fewer collisions in a slotted Aloha system of metropolitan area dimensions than in other random access systems, but comparative performance in practice remains to be determined.

Bulk Data Transmission

What we have been describing are "public" communication services. A very different approach is taken for dedicated links, such as private lines between locations of a business user. These are usually offered on the institutional cable, or a special private communications hookup, separate from the noise and interference problems of the residential cable. The arrangement is typically that of Fig. 2.38, with simultaneous two-way (full duplex) transmission provided by dedicating two upstream and two downstream channels to a user pair. As in a random access system, transmissions are relayed through the cable headend. Facilities at T1 (1.544 Mbits/s) rates and higher can be provided, usually at prices competitive with comparable facilities in the telephone network, although the costs of implementing reliable facilities have dissuaded most cable operations from doing so. Data rates in the hundreds of megabits per second are becoming feasible on the optical fiber systems introduced earlier in this chapter, which can be specially built for large commercial users.

The public and private facilities are not mutually exclusive. If more switching and a multiplicity of data channels can be provided in cable systems, and various signal level and noise problems resolved, several communications alternatives can be accommodated at once. A number of channels might be operated in a polling or random access mode for randomly generated light traffic, including requests for more permanent

Fig. 2.38: *Two-way*
point-to-point data communication
through private channels on an
institutional cable.

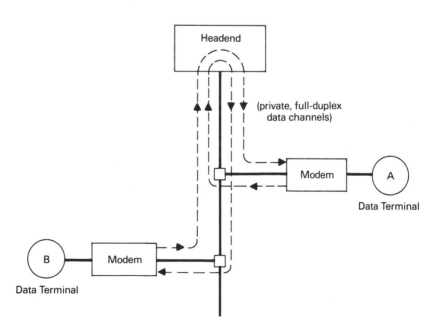

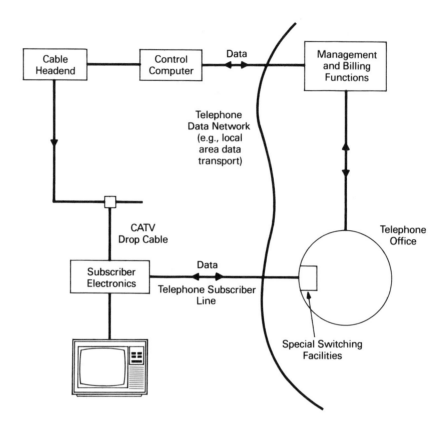

Fig. 2.39: Interactive cable system built as a hybrid of cable (downstream video and high-speed data) and telephone network (lower speed information and control signalling) facilities. By using special switching and data transport facilities, communication can be as fast as in a cable-only interactive system.

channel capacity, while others may be dedicated in a demand-assigned way. Still others can be completely reserved for private users. The channels can be defined as either separate frequency channels, or separate portions of the transmission time of one or a few high-speed channels.

Cable networks, like the telephone network, will have to build a variety of data communications facilities to service a variety of customer needs. Cable systems have the substantial advantage of a wide-band medium largely paid for by CATV subscribers, and in some cases may be able to develop excellent data communications capabilities with only a modest additional capital investment. Some examples are described in Chapter 4.

Hybrid Interactive Systems

An entirely different approach to interactive communications is to use the telephone network for the two-way data communication functions and devote the cable entirely to downstream television and higher speed data functions (Fig. 2.39). The telephone network, after all, is designed for two-way communication, while cable networks are largely broadcast "trees" which are intrinsically ill-suited to two-way communication.

Hybrid systems of this kind have many potential advantages. In addition to avoiding costly implementations of two-way communications in cable systems, they can realize economies of scale by sharing data communications networks with other customers. The local area data transport (LADT, see Chapter 12) networks being developed by telephone companies are attractive for this application. Sometimes billing functions can also be economically handled by the telephone companies operating the data networks, or by third parties connected through these networks.

Hybrid interactive systems can make it easy for cable subscribers to place requests and orders with service providers outside of the cable system, and participate in electronic mail and other new services of the outside data networks. They can also support important in-house services, such as impulse buying of pay-per-view programming. Without interactive data communications, but assuming, of course, addressable converters in which channels can be turned on or off from the headend, pay-per-view orders can be placed by telephone. However, the calling congestion can be intolerable just before a program begins, so that automatic and fast ordering through an interactive data terminal is highly desirable. A compromise technique, using ordinary telephones and a new telephone network capability for automatic identification of calling numbers, has been packaged by Zenith Electronics in its "Phonevision" system. The number identified by the telephone network as a caller to the pay-per-view ordering number is passed to the cable operator, who checks it for authorization and switches on the pay channel through the completely separate cable system.

There are both technical and business difficulties in realizing hybrid systems, some of which boil down to the question of control by cable operators of their communications and business operations. The economies and service advantages of hybrid systems should, however, encourage future development.

Cable Carriage of Enhanced Television and Other Special Signals

High-fidelity stereo sound has finally come to television (Appendix 1), and will be demanded by cable subscribers. During the development of an acceptable stereo system, cable operators were uneasy about the consequences for cable transmission. Effects such as an increase in harmonic and intermodulation distortions, incompatibility with scrambling formats and signal damage in baseband converters were noted. It was still not clear, at the time of this writing, if separate audio transmission channels would be necessary, requiring new subscriber equipment to receive the stereo sound. This is a striking example of how cable transmission can be more difficult and restrictive than over-the-air broadcasting.

Digital music (see Chapter 4) is another special signal which may be delivered on a large scale by cable in the future. As digital optical compact disks come to replace phonograph records, a market will develop for electronic delivery of digital recordings. A data broadcasting—not interactive—channel of several megabits per second capacity will be required for each program of this service.

Although no subscriber equipment or agreed-upon standards were available at the time of this writing, high definition television (HDTV) is the largest potential enhancement of television since color. HDTV usually means television with twice the standard number of horizontal traces (perhaps 1125 instead of the 525 of normal television in the U.S.) and twice the resolution along each line, plus a screen aspect ratio (ratio of width to height) of 5:3, closer to motion picture standards than the current 4:3.

The picture could have more than four times the normal content, a difference perceived as a much greater sharpness. This would not necessarily require four times the present television bandwidth, since the NTSC encoding is far from optimal; some proposed HDTV standards demand less than twice the present 6 MHz television bandwidth. HDTV is most impressive in projection viewing systems, where its clarity is comparable to that of a good motion picture film. The future home viewer may be able to view HDTV on a large, flat wall screen or on the wall itself as an image from a compact projection unit.

Enhanced definition television, in which large subjective improvements are possible through better transmission and signal processing techniques without going to a new display standard, could become available in an earlier time frame. Digital television receivers can, for example, interpolate new picture lines between transmitted picture lines, creating the impression of higher resolution.

Operators of future direct broadcast satellites (see Chapter 10) have publicly announced their intention to offer enhanced or high definition television at some future date, so that cable systems may have to accommodate similar signals if only to meet this competition. Cable has both advantages and disadvantages. On the positive side, the newer cable systems can easily provide the two to four standard video channels required for high definition television signals, and interactive capabilities are not required. But the sensitivity of the new very wide-band signals to the intermodulation distortions and other peculiarities of cable transmission remains to be determined.

What Else?

An attempt has been made, in the preceding pages, to describe the basic technology which supports present and future cable services. There are many other technical considerations, of considerable importance in building and operating a cable system, which have received little or no attention in this brief review. These include headend and studio design and construction, network layout, signal level and noise margin engineering, microwave or light-wave feed link construction, and all the other elements of building a system. Network layout is particularly critical because it must anticipate future demand with the minimal current investment in facilities.

After the transmission system is constructed, the operation must be supplied with the video and audio equipment associated with television work: cameras, lights, microphones, recorders, editors, test equipment, mobile video units, and so on. And finally, there are the technologies associated with business operations and auxiliary services. These include effective network control and billing systems, subscriber electronics packages, systems and applications software, and communications interfaces to the rest of the world. Detailed cable engineering is beyond the scope of this book, but the bibliography at the end of this chapter includes references which will supply additional technical information.

For all the newness and potential for developing into a networking medium, CATV is largely a simple technology for broadcasting a collection of video signals through a pipeline rather than through free space. It produces economies of scale by integrating a collection of signals which most individuals could not afford to do, and it promises, and often does deliver high-quality signals to subscriber locations where over-the-air signals may not be clearly received. These features are its distinct and almost unique reasons for being, although ancillary services may have a growing impact in the future.

Of course, these video delivery features may not be distinctive enough where many good-quality over-the-air signals, including pay television, are available; where a user group, such as the inhabitants of an apartment complex, can get less expensive service from a private cable system; where videocassettes meet people's needs for video entertainment; or where the telephone network provides attractive data-based services. These alternatives are described at length in the second part of this book.

For Further Reading

[1]   *A Cable Primer.*   Washington, DC: National Cable Television Association, 1981.
[2]   T.F. Baldwin and F.S. McVoy, *Cable Communications.*   Englewood Cliffs, NJ: Prentice-Hall, 1983.
[3]   E. S. Kohn, "Scrambling and cable-ready TV receivers," *IEEE Trans. Consum. Electron.*, vol. CE-28, Aug. 1982.
[4]   D.C. Coll and K.E. Hancock, "A review of cable television: The urban distribution of broadband visual signals," *Proc. IEEE*, vol. 73, pp. 773–788, Apr. 1985.
[5]   R.J. Hoss and F. R. McDevitt, "Fiber optic video supertrunking: FM vs. digital transmission," in *1984 NCTA Convention Record.*
[6]   M.D. Stahlman, "Bypass technologies for metropolitan area networks," in *Proc. 1984 IEEE Int. Conf. Commun.*, Amsterdam, May, 1984.
[7]   T. Sato and S. Sakashita, "High performance CATV converter," in *Proc. 1984 IEEE Int. Conf. Consum. Electron.*, June 1984.
[8]   M.L. Ellis *et al.*, "INDAX—An operational interactive cabletext system," *IEEE J. Select. Areas Commun.*, pp. 285–294, Feb. 1983.
[9]   R.M. Metcalf and D.R. Boggs, "Ethernet: Distributed packet switching for local computer networks," *Commun. ACM*, vol. 19, July 1976.
[10] N. Maxemchuk and A.N. Netravali, "Voice and data on a CATV network," *IEEE J. Select. Areas Commun.*, Mar. 1985.

3 Programming

The desire for better signal reception started the cable industry, but programming built it into the giant it has become. In metropolitan areas, where a number of television stations are available and good reception is usually not a problem, the promise of this many-channeled medium has always been the delivery of programming of a variety, interest, and entertainment value not found in broadcast television.

Although many would claim that the promises of rich, diverse, and artistically superior programming have not been kept, except in a few instances (Fig. 3.1), the entertainment demands of a large number of Americans appear to have been satisfied. Cable programming is a competitive business, and like other mass-media entertainment industries has tended to consolidate to a few companies and to gravitate toward material appealing to large audiences. Many of the program suppliers who tried to do something different went out of business, and only in the mid-1980's have a number of the surviving program services become profitable. "It looks like we will have about a dozen broad-based services that are viable, including about a half-dozen advertiser supported services," said John Reidy, a securities analyst, in 1984. "It won't be the old cornucopia of services that people once spoke about. But we know these should survive" [1]. (See the "For Further Reading" section at the end of this chapter for numbered references.)

The mass audience wants sports, news, popular music, and, above all, uncut, uninterrupted movies. Significant segments of this audience are interested in religious programming, sexually-oriented programming, health and "lifestyle" features, and ethnic programming. Delivering all of this profitably to its audiences is the goal of cable system operators, but they are only the retailers in a three-part chain. The first two parts are the *producers*, such as motion picture studios, and the *programmers*, responsible for packaging and distributing video services to cable systems, usually via satellite. The distinctions are not always clear. Many cable operators are also programmers, and the programmers have frequently jumped into production.

Satellite distribution, with its instantaneously created national broadcasting networks, has made most of the cable programming industry possible. It was not so easy, in the pre-satellite era, to create a nationwide television network, and that is part of the reason there were so few of them. The emergence of dozens of new networks and of their local distribution through

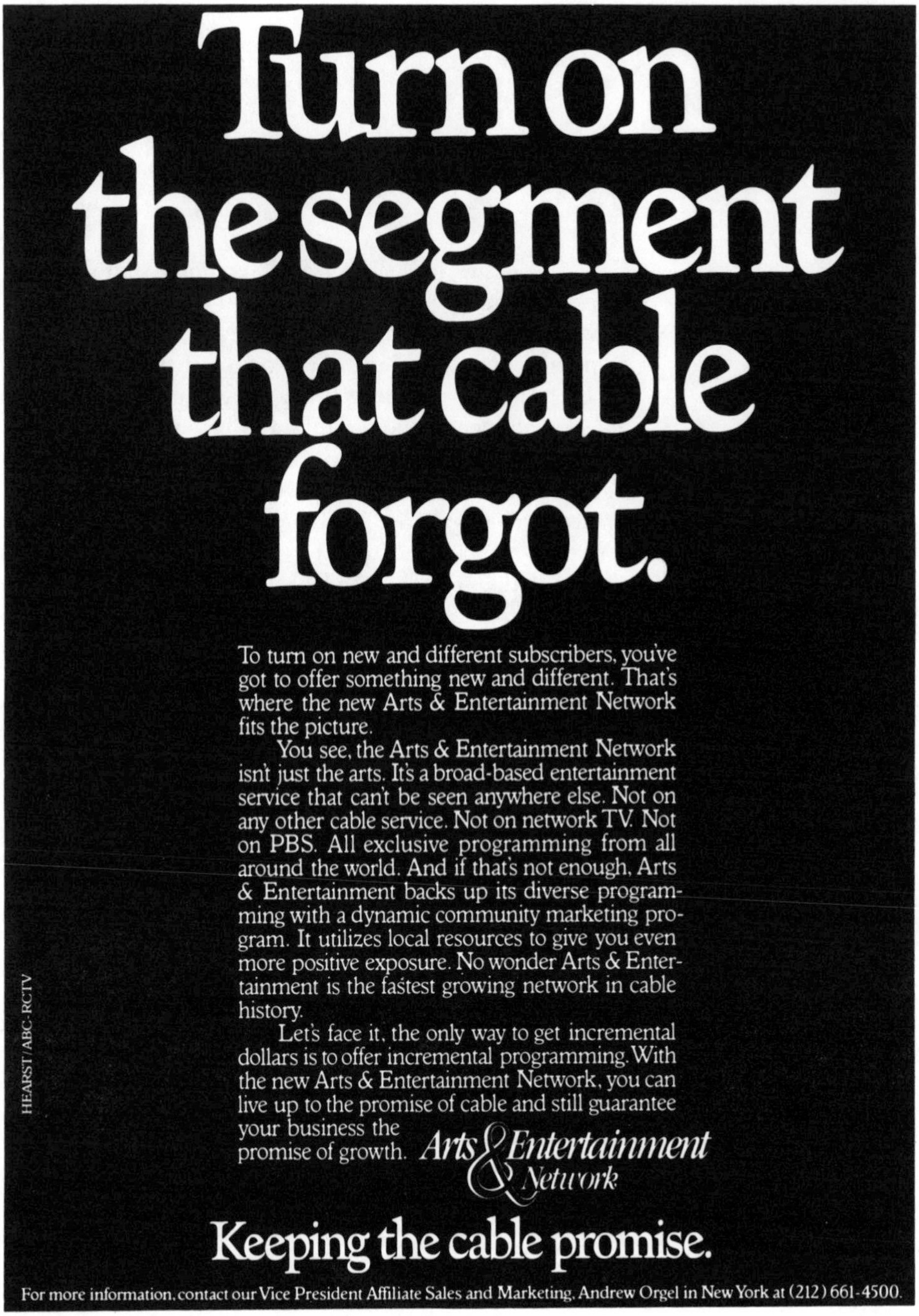

Fig. 3.1: An advertisement for a "cultural" channel. (Courtesy Arts & Entertainment Network.)

cable systems is one of the most striking consequences of the successful development of satellite communications and is certain to have long-range consequences for our individual and social lives.

This chapter offers a sketch of the programming industry, mainly through an overview of the program services, many of them national networks, which it distributes. It should be kept in mind that programming produced by the cable programming industry is not the only material carried by cable systems. It is augmented by local broadcast signals, locally generated programming such as high school sports and other community-specific features, and additional services, such as a program guide, produced by the cable operator.

The roster of program services generated by the cable programming industry is constantly changing, but the snapshot in time contained in this chapter should give a feeling for the mass entertainment industry which cable distribution has become. Perhaps it will also suggest the directions cable programming could take, including the long-awaited possibilities for something different from and better than broadcast television.

Programming Tiers

The evolution of cable programming has been fairly predictable. Early community antenna systems did not contribute anything extra, but merely repeated the programming offered by nearby broadcasters. As time went by, system operators began to offer programs no longer on prime broadcast television, such as reruns of broadcast programs and old movies, in order to fill the unused channels among the 12 VHF channels carried by the cable. A flat monthly fee was charged for the entire service.

As cable program services developed in response to a latent subscriber demand for something beyond this limited selection, in particular uninterrupted movies of more recent vintage, cable system operators began to group programs into *tiers*, each offered for a monthly fee. The first tier was, and is, the basic cable service, consisting of whatever channels and services are offered for the monthly connection fee negotiated with the municipal authorities. This tier includes all the local broadcast stations, as required until 1985 by FCC "must carry" rules (see Chapter 5). It usually takes in the other "free" programming described below, but a 1984 FCC decision limiting the regulatory authority of municipal and state governments encouraged some cable operators to remove these extra channels from the basic tier and put them in a new upper-basic tier for which an additional fee is charged. The Cable Act of 1984 eliminates, as of January 1987, all cable programming price regulation in areas where there is "effective competition," a term which the FCC must define.

Programming in the basic tier or tiers typically includes local broadcast stations, both VHF and UHF; "superstations," which are distant broadcast stations brought in by satellite; one or more news channels (Fig. 3.2); perhaps a financial news channel and a weather channel; sports-oriented

Fig. 3.2: Cable has pioneered the 24-hour all-news channel. (Courtesy CableVision Magazine.)

channels; one or more music "video" channels; a children's programming service; religious channels; a possible country music and entertainment channel; variety programming channels; and one or more ethnic programming channels. Some of these, such as Turner Broadcasting's Cable News Network (CNN) and CNN Headline News, are as professional as anything seen on the major broadcast networks. Music videos are short films, made by record companies or small professional video producers, in which the pictures accompany the music (mostly rock) rather than the other way around. Music Television (MTV), which originated with Warner Amex and is distributed by MTV Networks, was the first such service and for a long time the only full-time music video channel. It has been joined by Video Hits One (VH-1), intended for viewers over 25 and also distributed by MTV Networks.

Educational, professional, and public service channels, with the notable exception of C-SPAN, are hard to find in most cable systems. Broadcast from the House of Representatives and other sites in Washington, C-SPAN provides continual coverage of the legislative process, especially committee hearings, and has made stars of obscure politicians. In the larger systems, local politicians, high school sports, and people engaged in various community activities might appear on one or more community affairs or public access channels. When these are run by the cable operator, the production quality is tolerable, but when they are left entirely to the users, it can be terrible. The low level of production quality, in comparison with professional work, discourages viewer interest.

Text channels are quite common for channel schedules, sports and business news, and community happenings. As noted in Chapter 2, these occupy entire video channels, but there is interest in "teletext" brought in piggybacked on television signals as vertical blanking interval (VBI) data streams (Appendix 1). Music channels (sound only) are sometimes available, although not yet in the high-speed digital form, described in the next chapter, which would produce extremely high fidelity sound equivalent to that from compact disk recordings. More attention will be given to quality sound as television broadcasting increasingly moves to stereo and stereo television sets flood the market.

Cultural program services have not proliferated as many thought they would on multichannel systems. Despite the fairly large audiences which can be generated nationwide for classical music, dance, operas, and plays, individual cable systems have not found this programming as helpful in drawing subscribers as the other kinds of programming described above, and the cable operator must pay the costs in both channel occupancy and (in most cases) fees to the distributors. The production costs are high, and it is difficult to charge fees which are high enough to cover them. CBS abandoned its CBS cable cultural channel in September 1982, after just 13 months of operation. The only nationwide cultural channel in the basic tier at the time of writing was the advertiser-supported Arts and Entertainment Network (Fig. 3.1), which was doing well with an expected 12 million subscribers by the end of 1984.

Most of the basic program services, defined as "made for cable" services and offered, along with broadcast stations and local access channels, in the basic tier, are supported by fees paid to the program distributors by the cable operator and by advertising. According to a Nielsen survey, two-thirds of viewing of cable programming is of advertiser-supported channels. For years, neither source of revenue was adequate, and in 1982 and 1983 the 15 top cable channels lost an estimated total of more than $375 million. However, a "critical mass of programmers, viewers and advertisers," in the words of industry analyst Alan Cole-Ford, appeared to have been reached in 1984, and the picture turned dramatically better. Advertising has been the key factor, with advertisers attracted by the growing audience sizes and favorable demographics. A major variety service, the USA Network, tells advertisers "...we run our network like a network...we program every hour of the day to the largest available audience...you can pick the demographics and dayparts you do want without paying a price you don't want" (Fig. 3.3).

Cable advertising revenues rose to more than $500 million in 1984, a 15-fold increase over 1979. Most large television advertisers also advertise on cable program services, and the expectation is for increasing reliance by basic services on advertising and less on fees from cable operators. How this will affect the quality of programming, and its differentiation from broadcast television, remains to be seen.

The clustering and regional interconnections of cable systems described in Chapter 1 is important for *local* advertising revenues because it provides the

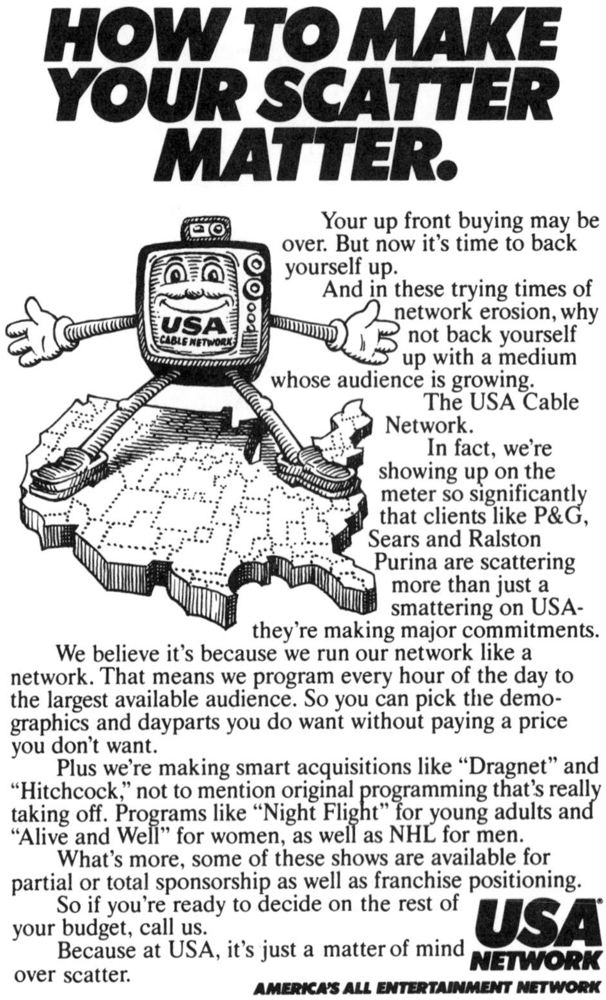

HOW TO MAKE YOUR SCATTER MATTER.

Your up front buying may be over. But now it's time to back yourself up.

And in these trying times of network erosion, why not back yourself up with a medium whose audience is growing.

The USA Cable Network.

In fact, we're showing up on the meter so significantly that clients like P&G, Sears and Ralston Purina are scattering more than just a smattering on USA- they're making major commitments.

We believe it's because we run our network like a network. That means we program every hour of the day to the largest available audience. So you can pick the demographics and dayparts you do want without paying a price you don't want.

Plus we're making smart acquisitions like "Dragnet" and "Hitchcock," not to mention original programming that's really taking off. Programs like "Night Flight" for young adults and "Alive and Well" for women, as well as NHL for men.

What's more, some of these shows are available for partial or total sponsorship as well as franchise positioning.

So if you're ready to decide on the rest of your budget, call us.

Because at USA, it's just a matter of mind over scatter.

USA NETWORK
AMERICA'S ALL ENTERTAINMENT NETWORK

Fig. 3.3: A USA Network advertisement in The New York Times directed to advertisers. Reprinted with permission of the USA Network®.

larger audiences which advertisers would otherwise find only on national program services. Ad interconnects are service organizations, sometimes but not necessarily run by one of the participating cable operators, which represent the participants to advertisers and distribute advertising to the participants. The interconnections are usually made by microwave or cable, but "soft" interconnects, defined as simultaneous running of taped advertisements on several systems, are also used, at least in the early stages. In 1984, there were 23 regional interconnects, each with an average of 9.5 cable systems and 1.9 million subscribers. The number of interconnects increased to more than 50 in 1985, with interconnect operators expecting revenue increases of 20–25 percent in 1985.

Advertisers and program distributors are so confident about the acceptance of advertising on cable that they are considering program services which are nothing but advertising. The Cableshop, a 24-hour channel carrying a mix of on-brand-specific informational and instructional advertising was scheduled to be launched in 1985. This is in line with the concept of an "infomercial," a long and informative commercial intended for potential buyers who are already interested but need more information before making a purchase.

The second tier conventionally contains satellite-delivered pay programming channels. Higher level tiers may be defined as combinations of pay channels or of pay channels and other satellite services.

Some pay channels are almost exclusively movies, and the movie production and distribution chain is by far the most significant part of the programming industry. It is described in a separate section below. Other channels, including some of the biggest ones, are predominantly movies but provide other entertainment as well, including live shows. There is even a return to soap operas, perhaps a little more daring than their television cousins, but recognizable just the same.

The movie and movie-entertainment channels are designed for general audiences, although more adult entertainment may be offered at late hours. Sexually-oriented services such as The Playboy Channel are obviously directed toward adults, especially men. They have been vigorously opposed by community groups unhappy with this kind of programming, but neither the law nor public opinion have favored censorship of scrambled pay programming. This is one area where there does seem to be a clear distinction between broadcast television and a scrambled subscription video channel.

There are also pay channels which feature sports, the arts (far fewer), business information delivered as text, electronic games, and personal computer software (see Chapter 4). Regional sports networks, featuring local teams, are a rising phenomenon. SportsChannel, in the New York area, broadcasts New York Yankee baseball and Islander hockey games, and the A-B (Anheuser-Busch) Sports Network in the midwest features St. Louis Cardinals, Cincinnati Reds, and Kansas City Royals baseball games as well as many other regional sports. These networks are supported partly by advertising.

The subscriber is assessed a monthly fee, usually $10 to $15, for any of these pay channels, in addition to the basic subscription fee. About half of this fee is passed through by the cable operator to the program supplier. Although eternally optimistic about the prospects for multiple pay subscriptions, cable operators have been disappointed by the poor subscriber response. A 1983 study by Benton and Bowles, an advertising agency, found that 17 percent of cable subscribers have cancelled at least one pay service for reasons other than changing residence, and only 10 percent of subscribers bought three pay channels at once. Only 0.8–0.9 pay tiers are sold for each basic subscription, and even the big new cable systems do not expect to take this ratio above 1.5.

Operators are still searching for the right pricing formulas, and may become more clever at assembling programming packages which will raise the ratio of pay to basic subscriptions, but they appear to be facing a saturation problem. There is just so much television that a paying adult will want, or have the time, to watch, and "home video" via videocassette recorders is taking a heavy toll. "Narrowcasting" programming to smaller audiences with special tastes, such as those who want foreign movies, folk music, or the classical performing arts, does sell additional pay subscriptions, but the operator's incremental costs tend, as already noted, to outweigh the incremental gains.

Because of the high costs and unexpected consumer resistance, the program distribution industry has been going through a period of consolidation, as a few prominent examples will illustrate. In 1984, ABC bought an 85 percent share of the Entertainment and Sports Programming Network (ESPN), an advertising-supported basic service, from Texaco for $202 million. It already owned the other 15 percent. Showtime, a pay service founded in 1978 as a joint venture between Group W and Viacom, went through several ownership changes. Viacom bought out Group W's share for $75 million in 1982, and in 1983 merged Showtime with Warner-Amex's The Movie Channel. An effort to include three movie studios, MCA, Paramount, and Warner Brothers, in the new venture was discouraged by the U.S. Justice Department. Although the two services were, when this was written, still run as separate entities, they were sharing management and satellite transponders.

As further examples, Spotlight, a pay service owned by several major cable operators, sold out to Showtime/The Movie Channel in 1984 and ceased operation. ABC and Westinghouse in 1983 sold their jointly owned Satellite News Channel to Turner Broadcasting's Cable News Network for $25 million, and the channel was discontinued. The Entertainment Channel, a pay service owned by RCA and Rockefeller Center Inc. ended operations in 1983 after large losses and little subscriber penetration. Cable Health Network merged with Hearst/ABC's Daytime service in 1984 to form the new "Lifetime" basic cable channel.

The mid-1984 survivors among the major pay services are described in Table 3.1. Further consolidations are likely to have occurred by the time this is read. The situation in film production and distribution, described later, has been even more turbulent.

Aside from some of the regional networks, virtually all of the major program services, both basic and pay, are distributed by satellite. Many of them are broadcast, with a time differential, on two separate transponders to the eastern and western halves of the country, so that subscribers everywhere can see the prime time programming during their local prime time. A cable operator may soon need a set of antennas which can view as many as ten C-band satellites, in orbit or planned to be launched, in order to be able to select from all of the available programming.

Table 3.1: Major pay services, and their owners, in 1984. Most of the owners also own cable systems and advertiser-supported program services.

Service	Owner(s)
Home Box Office (HBO)	Time, Inc. (parent of ATC)
Showtime	Viacom, Warner, American Express
The Movie Channel	Viacom, Warner, American Express
Cinemax	Time, Inc.
The Disney Channel	Disney Productions, Inc.
The Playboy Channel	Playboy Enterprises
Home Theater Network (HTN)	Group W
Bravo	Rainbow Programming
GalaVision	Spanish International Network (SIN)

Table 3.2 gives a capsule overview of all of the major satellite-delivered program services, pay and "free." The subscriber estimates (early 1984, with some modifications from 1985) for the pay services represent actual paying subscribers, but those for non-pay services are the total subscriber populations on the cable systems carrying them and do not necessarily represent viewers. Many smaller services have been left out; over 100 of them, including text and audio services, were available in 1984. Although satellite delivery accounts for the bulk of non-broadcast station viewing, it should be remembered that cable households spend a very large part of their viewing time with the broadcast television stations carried on the cable.

Movies

Despite the range of programming which is evident from Table 3.2, full-length movies remain at the heart of cable programming. Certainly they account for most of the money which is spent. The cable industry in the early 1970's was of little concern to the major motion picture studios, which regarded it as insignificant. But the cable industry developed an insatiable demand for movies, and the pay programming services have gone to great lengths, with and without the cooperation of the big studios, to obtain enough new material to meet this demand. Pay television may already be bringing in more revenue than theater box offices, and is profoundly changing the structure of the film industry. As Robert Lindsey of the New York Times wrote in 1983,

In an amazingly brief time, the six studios ... that have dominated the production, financing and distribution of films in America for most of

Table 3.2: Major satellite-delivered programming services available to cable systems in mid-1984.*

PAY PROGRAMMING (MOVIES AND GENERAL ENTERTAINMENT)

		Schedule	Satellite	Subscribers
HBO (Home Box Office): Family oriented, considerable original material.	Movies, entertainment, specials	Daily 24 hrs	Satcom F3 Galaxy 1	12,500,000
Showtime: Broad range of movies, including late night adult entertainment. Affiliated with The Movie Channel.	Movies, specials	Daily 24 hrs	Satcom F3 (W) Galaxy 1 (E)	5,000,000
The Movie Channel: 24-hour all movie channel.	Movies only	Daily 24 hrs	Galaxy 1	3,200,000
Cinemax: "Second service" complement to HBO; structured into time blocks for different audiences.	Movies only	Daily 24 hrs	Satcom F3 Galaxy 1	2,500,000
The Disney Channel: Family programming, Disney movies.	Movies, entertainment	Daily 19 hrs	Westar 5 Galaxy 1	916,000
The Playboy Channel: A video *Playboy* magazine.	R-rated movies, adult game shows, etc.	Daily 8 PM–6 AM	Satcom F4	668,000
HTN (Home Theater Network): Family programming.	Movies, musical	Daily 4 PM–4 AM	Satcom F3	267,000
SelecTV: Emphasis on films, including late night adult movies.	Movies, specials sports	Daily 24 hrs	Comstar D4	200,000
Bravo: Cultural channel, stereo sound.	Movies, concerts, discussions, etc.	Mon–Fri 8 PM–6 AM Sat–Sun 5 PM–6 AM	Satcom F4	196,000
GalaVision: Spanish-language service.	Films, specials, sports	Mon–Fri 4 PM–4 AM Sat–Sun 11 PM–4 AM	Galaxy 1	120,000

NON-PAY SERVICES

Superstations	Schedule	Satellite	Subscribers
WTBS (Atlanta): Cable-oriented independent station. Family oriented. Sports, movies, syndication, news, specials.	24 hrs	Galaxy 1	34,000,000
WGN (Chicago, channel 9): Rounded format with movies, sports, specials and syndicated programming.	24 hrs	Satcom F3	14,700,000
WOR-TV (New York/New Jersey, channel 9): New York's independent station with sports programming, movies, and old TV shows.	24 hrs	Galaxy 1	4,522,000
WPIX: Independent New York variety station with family-oriented programming, movies, sports.	24 hrs	Satcom F4	900,000

*Source: SAT Guide, June 1984; The New Yorker, May 20, 1985.

76 GETTING THE PICTURE

Table 3.2: *Continued.*

Public Service Channels	Schedule	Satellite	Subscribers

C-SPAN:
    Sessions and committee hearings from U.S. House
    of Representatives, plus other coverage.
	24 hrs	Satcom F3
		Galaxy 1
			19,300,000

News, Financial, and Weather Services	Schedule	Satellite	Subscribers

Cable News Network (CNN):
    Round the clock news coverage, in-depth reporting,
    features, interviews, financial, and sports news.
	Daily
	24 hrs
		Satcom F3
		Galaxy 1
			30,600,000

CNN Headline News:
    Compact news headline service, reports each
    half hour.
	Daily
	24 hrs
		Satcom F3
		Galaxy 1
			13,600,000

BizNet:
    Economic, legislative, regulatory news from business
    perspective for members of U.S. Chamber of Commerce
    and subscribing cable systems and TV stations.
	Mon–Fri
	6 AM–1 PM
		Satcom F4	37,000,000

Financial News Network (FNN):
    Live business and financial news including
    stock and commodity market quotes.
	Mon–Fri
	7 AM–7 PM
		Satcom F3	17,500,000

Business Times: Two hours of business news and features.	Daily 6 AM–8 AM	Satcom F4	1,600,000
The Weather Channel: Live national and localized weather service.	Daily 24 hrs	Satcom F3	14,000,000

Dow Jones Cable News:
    A *text* business/financial/economics news
    service derived from *The Wall Street Journal*,
    Dow Jones News Service, and *Barron's*.
	Daily
	24 hrs
	(VBI on
	WTBS)
		Satcom F3	1,770,000

AP News Cable: News, weather, sports, etc. (a *text* service)	Daily 24 hrs (VBI on WTBS)	Satcom F3	4,700,000

Variety Services	Schedule	Satellite	Subscribers

USA Cable Network:
    Broad-based variety programming, with special
    segments for particular audiences.
	Daily
	24 hrs
		Satcom F3	28,100,000

Lifetime: Health, self-development, and related topics.	Daily 24 hrs	Satcom F3	21,800,000

MSN—The Information Channel:
    Consumer information, series on travel,
    business, products, health.
	Mon–Fri
	10 AM–1 PM
		Satcom F3	9,300,000

SPN (Satellite Programming Network):
    Movies, entertainment, recreation,
    other variety programming.
	Daily
	24 hrs
		Satcom F3	11,500,000

USA Cable Network:
    Broad-based variety programming, with
    special segments for the different audiences
    expected at different times of the day.
	Daily
	24 hrs
		Satcom F3	28,100,000

Pacific Cable Network:
    Variety of programming including theater,
    sports, call-in shows.
	Daily	Satcom F4	—

Table 3.2: *Continued.*

Religious	Schedule	Satellite	Subscribers
EWTN (Eternal Word Television): Catholic religious programming.	Daily 8 PM–12 AM	Satcom F3	1,900,000
ACTS Satellite Network: Southern Baptist "family Christian entertainment."	Daily 10 AM–2 AM	Spacenet 1	1,000,000
PTL ("Praise The Lord") Satellite Network: Interdenominational religious programming.	Daily 24 hrs	Satcom F3	10,000,000
NCN (National Christian Network): Multidenominational religious programming.	Daily 8 PM–3 PM	Satcom F4	1,200,000
TBN (Trinity Broadcasting Network): Variety religious programming, including live coverage of major religious events in the U.S.	Daily 24 hrs	Satcom F4	4,500,000
The University Network: "24 hour nonsecular programming."	Daily 24 hrs	Westar 5	20,000,000

Music	Schedule	Satellite	Subscribers
Country Music Television: Video stero country music.	Daily 24 hrs	Comstar D4	4,000,000
MTV (Music Television): Video "rock" channel (surrealistic film clips accompanying music.)	Daily 24 hrs	Satcom F3	24,600,000
The Nashville Network: Country-oriented entertainment, mostly music.	Daily 9 AM–3 AM	Galaxy 1	20,400,000

Ethnic (See also pay programming)	Schedule	Satellite	Subscribers
BET (Black Entertainment Television): Variety programming with black performers in dominant or leading roles.	Daily 8 AM–2 AM	Galaxy 1	8,000,000
National Jewish Television: Jewish news, cultural, and religious programming.	Sunday 1 PM–4 PM	Westar 5	2,900,000
SIN (Spanish International Network): Spanish-language variety channel.	Daily	Spacenet 1	28,100,000

Special Audience	Schedule	Satellite	Subscribers
The Silent Network: Entertainment and information programming in sign language and voice.	Thurs 6 PM–8 PM	Comstar D4	5,700,000
The Prime of Life Network: Programming targeted to the 45 and over audience.	Daily 4 PM–7 PM	Satcom F4	—

Children's Services	Schedule	Satellite	Subscribers
Nickelodeon: Quality children's programming, preschool through teenage.	Daily 7 AM–8 PM	Satcom F3	22,700,000

Table 3.2: *Continued.*

Sports	Schedule	Satellite	Subscribers
ESPN: Professional, college, and amateur sporting events, including college football and basketball, NBA basketball, USFL and CFL football, professional golf and tennis, etc.	Daily 24 hrs	Satcom F3 Comstar D4	30,200,000
Madison Square Garden Network: Live sporting events from Madison Square Garden.	Daily 7 PM–4 AM	Westar 5	1,600,000
The Meadows Racing Network: Nightly harness racing servicing ''call a bet'' wagering system in Pennsylvania.	Daily except Mon and Wed	Westar 5	285,000
USA Blackout Network: Programming to replace events affected by local blackout regulations.	Daily 24 hrs	Satcom F3	22,800,000

Pay Sports Channels	Schedule	Satellite	Subscribers
Home Team Sports: Regional network for Washington, DC, and nearby states.	Daily 24 hrs	Galaxy 1	200,000
New England Sports Network: Regional network featuring Boston Red Sox and Boston Bruins.	Daily 6:30 PM– 12 AM	Satcom F1R	—
Sports Time Cable Network: Regional network in South and Midwest featuring Cardinals, Reds, Royals.	Mon–Fri 7 PM–1 AM Sat and Sun 1 PM–1 AM	Satcom F3	—
Home Sports Entertainment: Regional network from Dallas featuring Astros, Rockets, Texas Rangers, Dallas Mavericks.	Daily 6:30 PM– 1 AM	Satcom F4	50,000

Cultural and Educational Sevices (Non-Pay)	Schedule	Satellite	Subscribers
Arts & Entertainment: Visual & performing arts, including Broadway plays and BBC comedy series.	Daily 8 PM–4 AM	Satcom F3	12,000,000
The Learning Channel: Career-oriented programs, college credit courses.	Daily 6 AM–4 PM	Satcom F3	4,500,000
MSN (Modern Satellite Network): Informational, documentaries.	—	Satcom F3	10,100,000

the last 75 years have lost control of a large measure of their business...Hollywood simply did not recognize the importance of a fundamental shift in the way many Americans are going to the movies nowadays—in their homes instead of a theater...Time, Inc. [distributor of the Home Box Office pay television channel] has become, by far, Hollywood's largest financier of movies. [2]

The story of the production and distribution of films for cable television is one of program services struggling to guarantee for themselves steady supplies of new films, preferably at the expense of competing services. It began with efforts to persuade the studios to produce material for cable, and in 1977, the first year of serious effort in this direction, Columbia Pictures and 20th Century Fox did in fact make more than 30 films for cable.

But pay TV distributors, dissatisfied with both the quantity of available material and the price, showed early interest in becoming producers themselves. In April of 1977 HBO produced and aired "Standing Room Only," a variety entertainment show. Showtime negotiated for Broadway musicals and for production of a revue called "Sports on Ice." The distributors also cultivated independent film producers, who were willing to agree to exclusive contracts. In May of 1983 HBO carried "The Terry Fox Story," an independent production which made history as the first major movie to premiere on cable rather than in theaters. HBO has a dominant role in Silver Screen Partners, a direct partnership production company which made a $125 million public stock offering in 1983 and negotiated with the British Broadcasting Corporation to jointly produce a number of feature films in Europe.

These efforts did not help relations with the major studios. In the early years of the pay-TV industry, program distributors were willing to accept non-exclusive licensing agreements and to share the revenues with the film studios at a per-subscriber rate. The rapid growth of the subscriber base and the growing influence of HBO changed this attitude and led to efforts to change the structure of the distribution agreements, including temporary abandonment of the major producers. HBO and other large distributors secured exclusive rights from independent producers, and generally refused to purchase cable rights to major studio films unless they, too, came with exclusives. The distributors also sought flat fee versus per-subscriber rate agreements with the film studios.

The distributors have generally won this struggle, subject to antitrust regulations of the U.S. Justice Department, and motion picture companies have had to accept not only their terms, but also their participation as production partners. The studios have reason to feel they have lost something, because when a movie is shown in a theater, the studio that distributes it generally keeps about 45 cents of each box-office dollar. But when HBO distributes it, the studio gets less than 20 cents of each dollar paid by an HBO subscriber for monthly service. Nevertheless, the studios and pay distributors appear to have settled into a period of long-term cooperation in movie production.

This takes several forms. One of them is a joint venture arrangement, exemplified by Tri Star Films, formed by Home Box Office, CBS, and Columbia Pictures, which received antitrust clearance in 1983. It produces films mainly for theatrical and pay cable release, presumably with exclusivity rights, but CBS has broadcast rights to some materials. The trend toward multi-media distribution, with cable playing a dominant role, is also seen in agreements by broadcasters to acquire products originally made for cable, as in an arrangement which Metromedia, Inc. made with HBO.

A more general production arrangement is a purchase contract, preferably exclusive, between a pay program service and a film studio. Paramount Pictures and Showtime/The Movie Channel signed a five-year agreement in late 1983 that gave the two pay channels exclusive pay TV rights to all of Paramount's theatrical motion pictures. The price paid to Paramount was estimated between $400 and $700 million. Home Box Office, after efforts to conclude a similar exclusive agreement, signed a six-year nonexclusive contract with Universal Studios, providing for HBO purchase of the bulk of Universal's movie production, for release on HBO and Cinemax. A second agreement with 20th Century Fox was said to be in preparation. The two agreements could provide HBO with 30 movies per year, considerably easing the constant pressure to find new material.

With CATV expanding in other countries as well as in the U.S., it is not surprising that the pay services and movie studios are forming overseas ventures also. HBO, Columbia Pictures, and 20th Century Fox in 1984 joined a British film company, Goldcrest Films and TGelevision Ltd., and a British electronics and entertainment company, Thorn EMI PLC, in a program service called Premiere for sale to cable systems in the U.K. Thorn was the dominant partner. Distribution was expanded later in the year to Germany, Switzerland, and Austria, with Showtime/The Movie Channel as an additional supplier participant and Beta Films G.m.b.H. of Munich as the dominant local partner.

Amidst all this deal-making among powerful companies, movie industry workers also have demanded a significant share of pay-TV revenues. During 1980 and 1981, the Screen Actors Guild and the Directors Guild demanded royalties from films made for pay TV, videocassettes, and videodisks. The actors and directors actually went on strike, and the major motion picture companies suspended pay TV production, driving the distributors even further toward direct financing of independent producers. The problem ended with the development of a formula which gave actors and other employees a proportional share of pay TV fees.

Pay-Per-View

The relationships among film producers, cable programmers, and cable operators are likely to change further. Despite the large deals with studios, and a number of premieres on cable, the major step to first-run distribution via CATV has not yet been taken.

For first run showings, the industry will have to turn to a pay-per-view system, which means separate payment for a single event, just as in purchasing a ticket at a box office. Pay services sold by monthly subscription simply cannot bring in as much first-run money to film producers as theatrical distribution.

CATV pay-per-view services exist, but they require special equipment and have operational problems. Ordering a movie usually means a telephone call to the cable operator, unless the system is one of the few which are interactive as described in Chapter 2. The usual rush of last-minute orders can overload the order-takers. Automatic telephone order systems help, but the best solution is the development of interactive CATV systems, possibly, as discussed in Chapter 2, in hybrid systems with telephone data networks, which could respond to a large number of simultaneous subscriber requests with negligible delay.

Delivery of the pay-per-view programming is best done in addressable converters (see Chapter 2), through which the pay-per-view event can be switched by a computer command from the cable headend. Lower cost addressable converters are becoming available, and large cable MSO's have become more interested in installing them. However, only six of the 38 million cable subscribers had addressable converters in mid-1985. A crude but simple alternative is the one-time disposable trap, a device attached between the converter and the television set to remove an interference sent with the special programming to scramble it. But the disposable trap costs about $5, and is not easily distributed.

Another requirement is the efficient delivery of pay-per-view programming to the cable operator, now done mostly by physically transporting tapes. First-run distribution of pay-per-view movies to a large number of cable operators requires satellite broadcasting, just as for other cable programming. And finally, more modest prices, comparable, for movies, to a theater ticket or a videocassette rental, are imperative for consumer acceptance.

Subscribers have not shown overwhelming enthusiasm for the intermittent pay-per-view services which have been offered, and sales totalled only $26 million in 1984. But the potential is large, as suggested in Fig. 3.4, and cable operators continue to show interest. A test called "FirsTicket," run in 1982–83 by ABC Video Enterprises and Cox Cable in Cox's San Diego and

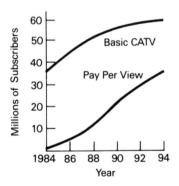

Fig. 3.4: Paul Kagan Associates, Inc. forecast of growth in number of pay-per-view subscribers. This research firm has predicted $1.5 billion in pay-per-view revenues in 1991. (Sources: The Wall Street Journal, Jan. 10, 1985; The New York Times, Aug. 11, 1985).

Santa Barbara cable systems, showed, according to Cox's president, David Van Valkenburg, that pay-per-view "can represent good incremental revenue without cannibalizing other revenues from premium service offerings" [3].

Three pay-per-view distributors, EvenTelevision, Caesar's World, and Video Techniques, were in existence in 1984. In early 1985, EvenTelevision, a joint venture among Group W Cable, Warner Amex Cable, ATC, Tele-Communications, Inc., Newhouse Communications, and Caesar's World Productions, promised monthly showings beginning with a series of boxing events, but, later in the year, announced plans to switch to recently released movies, eight to ten each month, supplied by satellite to cable operators. Playboy began satellite distribution of "Playboy Private Ticket," a combination of original programming and movies, in July 1985. Showtime/ The Movie Channel formed a new pay-per-view venture in 1985, with movies to be shown the day they are released on videocassettes, and was planning an end-of-year startup when this book went to press. Other pay-per-view distributors in 1985 were The Exchange, Choice Channel, and Megacom Telecommunications.

There are many alternative strategies for running pay-per-view programming. The usual concept is a one-time scheduled event, but the ultimate in subscriber convenience would be video on demand, a private movie showing at any desired time. A reasonable compromise might be a schedule of a number of showings of a certain movie during a week, with a subscriber permitted, for a fee of $5 or so, to tune in during any of the showings. Cable operators were thinking, by the mid-1980s, of a synergistic relationship with videocassette recorders in which these pay-per-view showings would become a more interesting source of program materials for home taping than the competing prerecorded videocassettes from rental stores.

It seems that cable is about to get a chance at first-run, or almost first-run, distribution of video entertainment. Perhaps pay-per-view will yet become a bonanza bringing in billions of dollars annually to distributors and cable operators.

The Outlook for Cable Programming

Can we look forward to different and better cable programming? Probably not in the near future. CATV distributors and operators want to become the preferred medium for mass entertainment, and attractive advertising vehicles for mass-marketed consumer products. They have no economic motivation, other than a marginal incentive to cable subscriptions, to build excellent channels narrowcast to relatively small and discerning audience segments. Cable operation is not as profitable as originally anticipated, and the industry is consequently shying away from programming innovation, even though it is anxious to build an image different from that of broadcast television.

The opportunity, however, is still there. If institutional support develops for narrowcast cultural and educational programming as it has for some religious and public affairs programming, the distribution capabilities of cable networks will indeed bring us something different and better. If interactive communication and pay-per-view were more widely available, they would add tremendously to the potential for more sophisticated entertainment, for effective professional and educational programming, and for greater use of CATV as a first run distribution network. A combination of factors is holding this back, but the largest factor is the lack of public and institutional will to sponsor and subsidize worthwhile programming which will never be able to pay its own way.

For Further Reading

[1] P. Kerr, "Cable TV's turn for the better," *The New York Times*, July 28, 1984.
[2] R. Lindsey, "Home Box Office moves in on Hollywood," *The New York Times Magazine*, June 12, 1983.
[3] *CableVision*, June 20, 1983.
[4] *Time Magazine*, Jan. 30, 1984.
[5] "Pay per view TV is gaining subscribers as fixed schedule cable loses favor," *Wall Street Journal*, Jan. 10, 1985.
[6] T. Whiteside, "Onward and upward with the arts: Cable 1," *The New Yorker*, May 20, 1985; for subsequent articles in the series see May 27, June 24 issues.

4 Ancillary Services

The wide-band cable which brings video entertainment to millions of viewers can supply dozens of other services as well. Cable can deliver information, education, security, electronic banking, and a string of personal and commercial communications services which go far beyond television programming. Like superior video programming, these are potentials of the cable medium which have been promised for decades but delivered only sporadically and in limited ways.

Trials of ancillary services have failed more often for lack of consumer interest than technical difficulties. It is not obvious that consumers will, in the near future, want such services at all, although it is hard to believe that some of them, if offered at the right price to the right customers, would not succeed. Most cable operators of the mid-1980's have seen little profit in ancillary services and have avoided them entirely, especially after the Cable Act of 1984 made it possible that many of them could be subject to state regulation.

Despite the overwhelming market viability questions, and the financial, technical, regulatory, and managerial obstacles, cable operators are likely to take another look at ancillary services. They cannot ignore cable's unique and persuasive advantages, especially its wide-band downstream capacity, for business communications and subscriber electronic services. There exists, in the 1980's and possibly into the 1990's, a window of opportunity for cable because of the limited transmission capacity of the telephone network at the subscriber level. This limitation may eventually be overcome, but, in the meantime, it offers cable technology and operators a chance to gain a foothold and important operating experience. Cable has the opportunity to become a communications medium of broader utility, and eventually to integrate into a general wide-band communications network.

This chapter introduces several of the many possibilities for ancillary services on cable, both residential and business-oriented, which could expand the role of the cable medium and move it toward the integrated communications environment suggested above. The services are introduced more or less in order of increasing communications capacity and sophistication.

Residential Services

The ancillary services can be grouped into two general classes: those directed to residential subscribers, and those intended more for commercial

Table 4.1: Some possible telecommunications services in the home, as envisioned by James Martin in *Future Developments in Telecommunications*. (Prentice-Hall, 1977, 2nd edition, with permission.)

Passive Entertainment
  Radio
  Many television channels
  Pay TV (e.g., Home Box Office)
  Dial-up music/sound library
  Dial-up movies
  Subscriber-originated programming

People-to-People Communications
  Telephone
  Videophones
  Still-picture phones
  Videoconferencing
  Telephone answering service
  Voicegram service
  Message sending service
  Telemedical services
  Psychiatric consultation
  Local ombudsman
  Access to elected officials

Interactive Television
  Interactive educational programs
  Interactive television games
  Quiz shows
  Advertising and sales
  Politics
  Television ratings
  Public opinion polls
  Debates on local issues
  Telemedical applications
  Interactive pornography
  Betting on horse races
  Gambling on other sports

Still-Picture Interaction
  Computer-assisted instruction
  Shopping
  Catalog displays
  Advertising and ordering
  Consumer reports
  Entertainment guide
  City information
  Obtaining transportation schedules

  Obtaining travel advice/maps

Computer Terminals
  Income tax preparation
  Recording of tax information
  Banking
  Domestic accounting
  Entertainment/sports reservations
  Restaurant reservations
  Travel planning and reservations
  Computer-assisted instruction
  Computation
  Investment comparison and
    analysis
  Investment monitoring
  Work at home
  Access to company files
  Information retrieval
  Library/literature/document
    searches
  Searching for goods to buy
  Shopping information; price lists
  Real estate searching
  Job searching
  Vocational counseling
  Obtaining insurance
  Obtaining licenses
  Medicare claims
  Medical diagnosis
  Emergency medical information
  Yellow pages
  Communications directory
    assistance
  Dictionary/glossary/thesaurus
  Address records
  Diary, appointments, reminders
  Message sending
  Christmas card/invitation lists
  Housing, health, welfare
  Games (e.g., chess)
  Computer dating
  Obtaining sports partners
  Obtaining travel advice and
    directions

  Tour information
  Boating/fishing information
  Sports reports
  Weather forecasts
  Hobby information
  Book/literature reviews
  Book library service
  Encyclopedia
  Politics
  Real Estate
  Games for children's
    entertainment
  Gambling games (bingo)

Monitoring
  Fire alarms on-line to fire service
  Burglar alarms on-line to police
  Remote control of heating and
    air conditioning
  Remote control of cooker
  Water, gas, and electricity
    meter reading
  Television audience counting

Telephone Voice-Answerback
  Stock market information
  Weather reports
  Sports information
  Banking
  Medical diagnosis
  Electronic voting

Home Printer
  Electronic delivery of newspaper/
    magazines
  Customized news service
  Stock market ticker
  Electronic mail
  Message delivery
  Text editing; report preparation
  Secretarial assistance
  Customized advertising
  Consumer guidance
  Information retrieval

users, although there is a heavy overlap. On the residential side, Table 4.1, from a book by James Martin, lists the many services which Martin conceives as being offered to home users. A large proportion of them can be delivered only on interactive systems with an upstream data channel, and in some cases an upstream video channel. Pay-per-view television (see Chapter 3), although not one of the ancillary services described here but important to future pay cable revenues, can benefit from the same interactive technology.

Table 4.2: Ancillary services which have attracted the interest of cable operators and equipment manufacturers (not in order of importance).

- Viewer response systems
- Security services (fire, burglar, medical alert)
- Environmental management (energy management, meter reading, appliance control)
- Games
- Microcomputer software
- Digital music
- Teletext (magazine-format information pages)
- Videotex (interactive, user-directed information and transaction services)
- Educational and social services
- Commercial communications services (business and residential, data and voice)
- Specialized business communication services

Many of the ancillary services listed in Table 4.1 are within the broader categories, listed in Table 4.2, which have received serious attention from cable operators. Investment capital has been scarce for reasons beyond doubts about market acceptance. "Cable companies are struggling under the huge capital burdens needed to meet their franchise commitments," according to a 1983 statement by Richard Neustadt of Kirkland and Ellis, a communications-oriented law firm, "and are facing long delays before their investments yield profits. Such companies seem unlikely to leap into the huge additional investments required for full-scale two-way services" [1]. (See the "For Further Reading" section at the end of this chapter for numbered references.) Just the same, some cable operators have been willing to make modest incremental investments to test the market for ancillary services.

Most ancillary services require interactive data communications. The Warner Amex Cable Communications Company, with the 1977 introduction in Columbus, Ohio, of its QUBE interactive system, was the first to provide a subscriber "talk back" capability. It was described by Gustave Hauser, then Warner Amex chairman, as "a supermarket of electronic services" [2], which it was not then and which no interactive cable system is yet. It was gradually extended to five other cities, reaching a total of over 300,000 subscribers, and was instrumental in winning some of the franchise competitions described in Chapter 5. But it was not an economic success, and interactive programming distributed to the six participating systems was drastically cut back in early 1984, after $30 million in losses and evidence that QUBE did not increase the number of cable subscribers. The interactive subscriber equipment was actually removed in some cities. *The New York Times* reported in early 1984 that only 500,000 homes nationwide had interactive cable and that it was unlikely to become widely available for many years. Trygve Myhren, the head of the ATC cable company, stated flatly that "Two way cable costs you more than it gets you" [4], an opinion which may not stand the test of time.

QUBE was used for a variety of experiments, from ordering pay-per-view programs to polls of viewers on political speeches and sports events (Fig. 4.1). It was a favorite with visionaries of the electronic future, and was

regarded for a time as the basis on which to build banking, shopping, and other more sophisticated services, but was never fully developed and did not succeed with viewers. "We failed to develop program forms which would make the passive television audience into active two-way participants because we patterned the programs after existing television,"said Michael Dann, who planned QUBE programs in the early days. "We did not create programming indigenous to the two-way system" [3]. QUBE, and systems like it, may survive in a limited form by concentrating on pay-per-view, as described in Chapter 3.

Limited as it was, the QUBE service attracted interest because it raised many political, social, and financial questions about two-way electronic services. Some critics felt that its "electronic democracy" features were potentially dangerous, since they did not provide either the universal access or stabilizing delays of the normal political process. Others suggested that viewer privacy was threatened by the system's computer records of what each subscriber was watching, although Warner Amex instituted a strict privacy code, and no significant breaches of privacy were reported. Small-scale experiments made with new forms of advertising, with pay-per-view, and with electronic shopping demonstrated the potentials of these services, if not the immediate practicality.

The major problem with the first generation of interactive systems, aside from cost, was, perhaps, not any threat to social stability or intrinsic consumer hostility, but simply their inadequacy for the full-scale information and communication services which would meet real human needs. If cable systems cannot offer something better or less expensive than what is possible through existing media, the telephone network, and long-established retailing systems, there is little justification for the investment in new equipment. Higher performance systems, offering much greater "on demand" capacity, may, if they can put together the right mix of services at acceptable cost, give cable operators a second chance to open up the interactive services market.

Security and Environmental Control

Fig. 4.2 suggests the generic design of a comprehensive cable-based home alarm and energy management system, using the same low-rate interactive data capability used for opinion polling and channel selection. The cable headend is in communication with an electronic box somewhere on the subscriber's property, a box which might also have a telephone line connection. Wires (or carrier current connections through the building's electrical wiring) run to heat, smoke, and entry sensors, the electric, gas, and water meters, appliances, and light switches. The cable headend sends instructions, such as "turn off the water heater," and regular polls, such as "have any of the entry sensors been activated?" The telephone line, when available, provides dialed backup to the interactive cable system. Alterna-

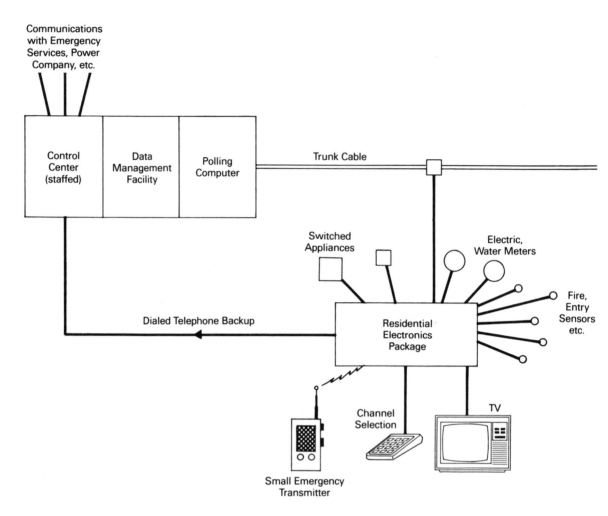

Fig. 4.2: Outline of a security and energy management system.

tively, as in a hybrid system called Transtext, announced in 1984 by Integrated Communication Systems, a telephone network data service can provide all of the lower rate data communications, eliminating the need for two-way cable, but presumably not for downstream data transmission for information services.

The security services tried thus far do not seem to have been a great success. It is not clear that the revenues support the operating expenses, even assuming the capital investment in an interactive system has been made for other reasons. Security systems will probably be profitable only as part of a package of interactive services sharing the capital investment and operating expenses.

The active polling suggested above provides warning of failure of the subscriber's security system which would not be provided by competing passive telephone line systems, which dial up a control center only when calling in an alarm. This capability for being constantly on line, not yet generally available to telephone subscribers, is one of the outstanding competitive attributes of interactive cable. It will not be an advantage for long, assuming telephone companies can overcome regulatory obstacles to providing efficient and low-cost data networking services (see Chapter 12) to residential subscribers.

An effective cable-based alarm system does not operate without human participation. A duty officer (Fig. 4.3) in a monitoring station receives alarms and attempts to telephone the affected subscribers before passing alarms to police, fire, and other public services. False alarms due to equipment malfunction and other reasons have been common. A subscriber may also initiate an alarm manually at a security system control box, or even

Fig. 4.3: *The subscriber's control console and the network control center in a cable-based security system originally developed by Warner Amex. (Photos courtesy Warner Amex Cable Communications.)*

from a small, personally carried transmitter which relays a signal to the control box. This is especially helpful in "medical alert" services provided to disabled or elderly persons.

Environmental management via cable is attractive to power companies for the possibility of reducing peak demand by switching off air conditioners, water heaters, and other heavy loads. As a general rule, the subscriber's comfort and safety must not be compromised in the interest of remote load management, and local manual overrides must always be available.

Microcomputer Software

Cable is an alternative to floppy disks and program cartridges for the delivery of recreational and other "software" to home computers and electronic game consoles (Fig. 4.4). Using special data channels set up in the cable for this purpose, paying customers can be given access to a considerable range and variety of computer programs. Even though the programs, when running, are interactive with the user, the delivery of software is not necessarily interactive. Programs can be simply broadcast, like non-stop movies, on a pay channel, with the subscriber's equipment "grabbing" a desired program as it is received in the transmission cycle.

The PlayCable channel, designed to supply games to Mattel Intellivision consoles, was made available to cable systems in 1981, but did not build a large customer base. PlayCable was delivered through a special data channel in the unused FM band (88–108 MHz) in the cable.

A more ambitious software broadcasting service, using a full video channel, was experimentally launched in the United States in 1984 by Nabu, a Canadian company which had been testing the system in Ottawa. Approximately 100 programs, including games, word processing, business analysis, and computer education, were continuously broadcast in a ten-second cycle. The system was designed to discourage software piracy through special hardware which prevented a subscriber from storing a captured program on a floppy disk; it could only be used in specially equipped microcomputers.

The broadcast delivery of games is only a small step toward the game possibilities on cable. With adequate *interactive* communication, players at different locations might play against each other or against distant computers in games of a sophistication, variety, and level of excitement which do not yet exist. This is a potential which may or may not be realized, but offers some hope for the evolution of video games beyond the arcade concept of an isolated playing machine.

Programming on Demand

There are, in fact, many electronic products which could be delivered to receiving devices, possibly in the dead of night, on channels which are busy

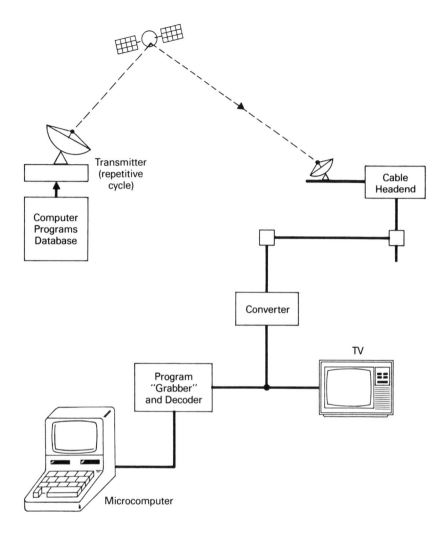

Fig. 4.4: Software distribution via satellite and cable. The "grabber" selects a desired program from a continuously transmitted selection of as many as several hundred programs.

with video programming during the day. Full-length motion pictures could be delivered to videocassette recorders of a subscribing audience, as ABC planned to do through television broadcast stations. This effort was abandoned in 1984 ostensibly because of competition from videocassette retail stores, but possibly also because of operational difficulties. Cox Cable Communications claimed more success with a service called World Video Library, in which subscribers schedule their VCR's to the hour nearest to when the desired event is to be shown. Analog music recordings could similarly be delivered to home tape recorders at off hours. In the future, digital sound will be carried by cable to digital playing and recording devices for the super high fidelity digital music described in the next section. These are all scheduled broadcasting services.

A long-standing dream of electronic services entrepreneurs has been to deliver programming on a one-time order basis to a single subscriber or a small group at the time specified by the subscriber. This places large demands on transmission capacity, since a number of subscribers may simultaneously want different programming. Programming on demand is not out of the question on the newest cable systems with more than 50 channels, but depends on the complexity of the programming material. Videotex services, with relatively low demands on communications capacity, are certainly feasible, and movies on demand are at least an intriguing possibility (Fig. 4.5).

A development effort to provide movies and other materials on individual subscriber demand was pursued by Warner Communications, but abandoned early in 1984. Hundreds of thousands, even millions of video channels would be required to meet the evening movies-on-demand needs of the population of a large city. It is much more likely that frequent scheduled broadcasting of a few hundred popular films, supplemented by a few video-on-demand channels for less frequently viewed films, will be the closest approach to true video-on-demand. Even this compromise demands more capacity than cable systems have today.

Fig. 4.5: *Full motion video on demand, with video segments stored on a bank of videodisk machines (or other memory media) at the cable headend.*

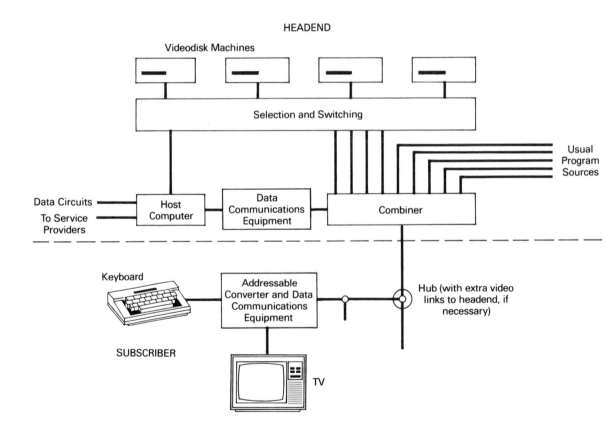

Better Sounds: Multichannel Audio and Digital Music

The relatively poor quality of the sound accompanying television signals has for many years concerned television engineers and program originators. A new industry standard for multichannel television sound (see Appendix 1), to be used for stereo or dual-language transmissions, will bring large improvements to broadcast television. However, stereo television sound may have to be translated to a separate audio carriage system on cable systems, which have technical difficulties in delivering stereo television signals.

A super high-fidelity *digital* music channel would be a completely different service, totally separate from television programming and without the technical problems of television sound. True digital recordings, which are now available on laser-scanned compact disks and may eventually be sold on tape cassettes, are an innovation in audio technology which is certain to become the standard for music reproduction. Just as with games and computer software, digital music can be delivered as a paying broadcast service on cable systems, either by monthly subscription or on a "pay-per-listen" basis. As Fig. 4.6 illustrates, all that the subscriber requires, in addition to existing high-fidelity equipment, is a receiver for the modulated signal carrying the digitized stereo sound and a digital-to-analog converter. These are relatively inexpensive devices. Several equipment manufacturers demonstrated experimental systems in 1984.

Only a wide-band transmission medium could accommodate this service. The information rate of perhaps two million bits per second required for high-quality digital stereo music calls for a transmission signal of considerable bandwidth. It would be relatively easy to send three to six stereo music signals in one video channel.

Teletext and Videotex

Information and transaction services have, perhaps, the largest potentials of any ancillary services on cable. The technologies of *teletext* (one-way broadcast) and *videotex* (two-way interactive) can be used to deliver consumer-oriented information "frames" by cable as well as through other media. Teletext and videotex are formally defined by standards, described in Appendix 2, for the generation, transmission, and presentation of frames such as those of Fig. 4.7, but are often loosely understood as, respectively, any broadcasting or interactive electronic information system.

The standard transmission medium for teletext is broadcast television, from which it gets a free ride in the largely unused vertical blanking interval of the television picture transmission cycle (Appendix 1). For videotex, the connection to an information provider is usually made through the telephone network. Although teletext frames are broadcast in a continuous cycle and videotex frames are transmitted only upon user request, they share presentation attributes and can be combined in a common facility which, to

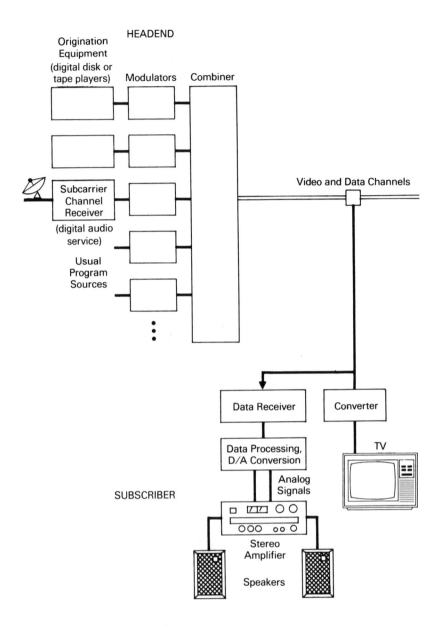

the user, appears to be a single information retrieval system. A cable-based information system could offer an efficient hybrid system in which a relatively small number of frequently requested frames would be continually broadcast to all subscribers, and a huge collection of less frequently requested frames would be available on subscriber request.

As ancillary services on cable, or, more properly, the vehicles for banking, shopping, electronic mail, and other transactional and information services,

Fig. 4.7: Typical teletext and videotex frames. (Courtesy Cox Cable Communications.)

teletext and videotex can go beyond matching the capabilities of television-carried teletext and telephone line videotex. In particular, data can be sent downstream at a much higher rate, and the circuit establishment delays of dial-up telephone line access to videotex services can be avoided. A travel or theater reservation system, for example, could be more effective on cable than on the dialed telephone network because of the absence of dialing delays and the faster display of information.

As described in Chapter 2, the cable medium allows a number of alternative ways to provide teletext and videotex capabilities. The simplest one-way system, which is extremely wasteful of transmission capacity, is to make a television signal from the teletext data stream at the cable headend, displaying each information frame for a minute or two and transmitting it in a full video channel, as a number of Southern California cable systems did during the 1984 Los Angeles Olympics. A second way is to do the same as television-borne teletext and use the vertical blanking interval of a video signal as the teletext channel (Fig. 4.8). A teletext magazine of 200 frames can be transmitted in about 12 seconds, and is constantly repeated. The subscriber's decoder has a "frame grabber," as in the software delivery system described earlier, which collects the roughly 1200 bytes of data defining a desired frame for local storage and display. There have been disputes between the cable and broadcasting industries over the right of a cable operator to strip away the teletext data on a TV signal picked up for cable transmission and substitute another, but the FCC has affirmed the right of the cable operator to do so.

The more interesting technique of full frame teletext, or some other high-rate data broadcasting system which does not adhere to the teletext standards, takes advantage of the high transmission capacity of cable distribution systems and can provide many times the performance or transmission capacity of a VBI teletext system. A system called "Request," which began field testing in 1984 on Group W Cable's Buena Park, California, system, offered 5000 pages.

Fig. 4.8: Teletext on cable, using the vertical blanking interval (VBI) of a video signal as the transmission channel.

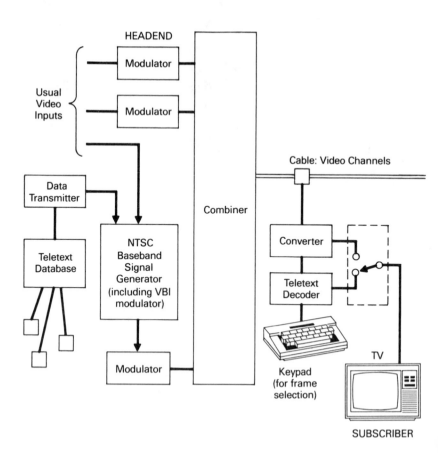

More important, perhaps, than the great quantity of teletext information which cable can accommodate is the opportunity for delivery of higher quality graphics without paying an unacceptable price in transmission time. These can go beyond refined graphics to animation and good quality still photographs, critical to educational, merchandising, and other applications. The commercial deployment of such enhanced teletext services is, however, contingent on the development of complex but inexpensive teletext decoders. Television set manufacturers are just beginning to build first-generation teletext decoders into their products.

Interactive videotex (Fig. 4.9), which supplies specific display frames in response to individual subscriber requests, can benefit even more from the high downstream transmission capacity of cable. Relatively low-speed upstream channels, sufficient for user requests, can be provided in the cable or through the telephone network in a hybrid system.

Fig. 4.9: Videotex on cable. A high-speed data channel, assigned on demand to a subscriber, delivers display frames, while a lower speed "upstream" channel carries requests and other information from the subscriber.

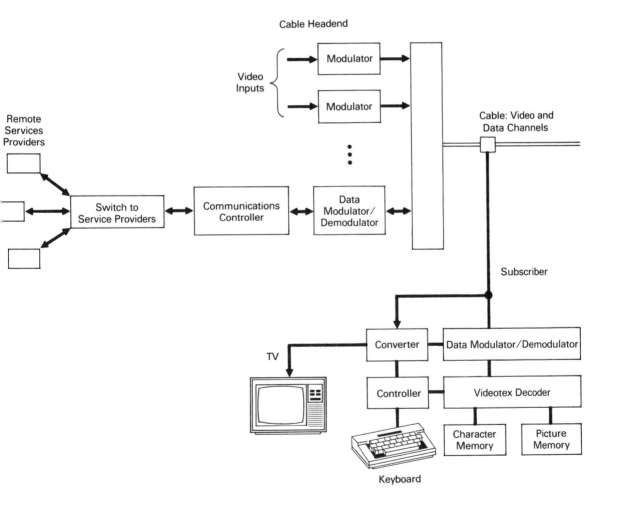

With a high-speed data channel, typically 18–256 kbits/second, assigned on demand to delivery of videotex frames to a particular user, a television screen can be filled with a normal text and graphics display in a fraction of a second rather than the 5–10 seconds of dial-up telephone line videotex. Photographic and other, more complex displays would take longer, but still be within the user's tolerance. This capability of cable for rapid downloading of information, and thus better quality screen presentations, is unmatched by telephone line videotex. The same display generation and memory components are used in teletext and videotex decoders, and can be shared by the two services.

It is hardly necessary to enumerate the services which these superior teletext and videotex capabilities could support. They include electronic banking and shopping, travel schedules and ticketing, stock market quotations, transactions and analytical software, personal messaging to other individuals, access to library indexes and reference services, and many of the others listed in Table 4.1. It is only fair to note that the technical and business difficulties of linking cable systems to a myriad of service providers are likely to be as significant as those associated with building teletext and videotex into cable systems.

Telephone Company Bypass for Communications Services

The cable industry's interest in commercial voice and data communications is partly its own idea and partly stimulated by outside events. Internally, cable operators have always wanted to make profitable use of spare capacity, particularly in the institutional cable. It should be possible to offer communication circuits at relatively low prices in a cable whose cost has already been absorbed as part of the cost of obtaining a franchise. The institutional cable runs past many businesses, is not plagued (or at least not as much) by the noise problems of the residential cable, and is usually not much used by the public institutions for which it is intended. Of course, there is not always an available institutional network, and upgrading costs can be very high.

The external factor, associated with an explosion in business communications and the deregulation of the U.S. telecommunications industry, is the interest of larger businesses and long-haul communications carriers in bypassing the local telephone company in order to lower local distribution costs. Telephone business tariffs can be unreasonably high because of regulatory constraints on relative pricing of commercial and residential telephone services. When bulk communications transport is called for, coaxial cable, optical fiber, and microwave are appropriate transmission media which are likely to be used whether or not cable operators are involved. Some cable operators are so attracted to this business that they are willing to go beyond their installed systems and compete for contracts to build new, dedicated communication facilities.

The usual requirements on the business side are for bulk transport of voice, data, and other services among business locations and from business locations to the serving offices of interexchange (long distance) communications carriers (Fig. 4.10), using the techniques described in Chapter 2. The traffic is frequently all digital, carried on cable data channels operated at digital rates of 1.544 Mbits/second and upward. A 1.544 Mbit/second "DS1" transmission system can carry 48 digitized voice channels in newer equipments. A company with several offices or factories in a metropolitan area can thus set up a substantial number of private lines among its locations, or a number of private line connections to a long-distance telephone carrier, at a cost which might be considerably less than the tariffed rates of the local telephone company.

Many examples could be given. The LINK Resources Corporation named seven cable companies with for-profit data services in early 1984, with estimated total revenues from these services of $7 million, forecast to grow to $103 million in 1989. Manhattan Cable, part of Time, Inc.'s ATC, began providing data transmission services to Bankers Trust in New York City in 1974. In 1984, it supplied more than 200 data circuits to commercial institutions, mostly banks, and to the City of New York. A specially constructed 17-mile underground business services network provides these circuits. Warner Amex, in 1984, was preparing to bid on construction of an internal communications system for the city of Dallas, to replace telephone company facilities. The intention was to operate the facilities as a "carrier's carrier," offering bulk capacity to various resellers. Cox Cable Communica-

Fig. 4.10: *Bulk business communications from business location to long distance telephone carrier, bypassing the local telephone company. PBX: private branch exchange, a business user's telephone switch. LAN: local area (data) network spanning a small area such as a building.*

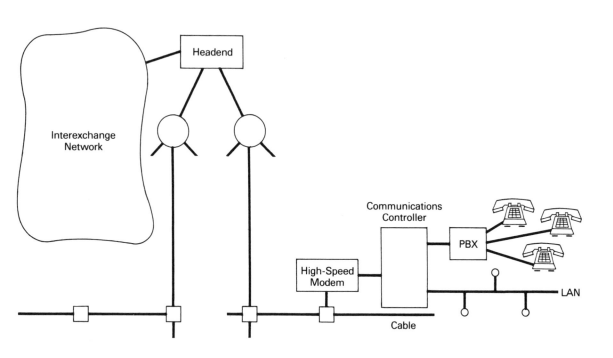

tions began an experiment in Omaha, Nebraska, in 1983 with data communications which included links between business users of communications services and interexchange communications carriers such as MCI and Sprint. It had plans to extend the service, called "CommLine," to other cities. MCI later broadened the idea to bulk voice communications and instituted a new service called "Cablephone," offered in at least five cities. Despite these examples, most cable operators, including Cox, have been pessimistic about the business potential of bypass communications. Cox, in fact, closed down its CommLine at the end of 1985.

The telephone operating companies which are the targets of this "limited" activity have challenged cable company intrusions on their traditional service areas. They have, for the present, won their demand that voice and data communications services on cable systems be subject, at the option of state public utilities commissions (PUC's), to the same state regulation as the comparable services in the telephone network, as described in the next chapter. This could put a damper on CATV exploration of voice and data services, although a cable operator convinced of the business potential of these services will not be stopped by regulatory red tape. Furthermore, state PUC's may require only informational tariff filings from CATV providers of these services, and the FCC may ultimately preempt state authority in this area.

Even for residential use, cable has some potential as a bypass medium. Voice communications can be provided through a collection of analog voice channels sent in separate narrow frequency channels, two to a user (one in each direction), connecting the user to a long distance communications company. This single-channel-per-carrier (SCPC) system might use a simple form of frequency modulation equipment. Unless effective scrambling schemes are devised and employed, any analog system of this type has the serious drawback of lack of privacy, since the cable, unlike a private telephone line, is accessible to many subscribers. MCI has considered extending its Cablephone service to residential subscribers, although it was initially offered only on institutional networks.

The architectural limitations of the tree structure of traditional cable systems call for special engineering efforts for bypass communications, some of which were suggested above, but there is still much to offer in the existing tree and branch networks. If cable operators emphasize bulk point-to-point transmission, broadcast applications, and applications requiring more capacity from a central site to peripheral locations than in the other direction, they can maximize their advantages. Videoconferencing is a good example.

Videoconferencing

A popular form of videoconferencing sends video of a speaker or event from a central studio to a number of remote viewing locations (Fig. 4.11).

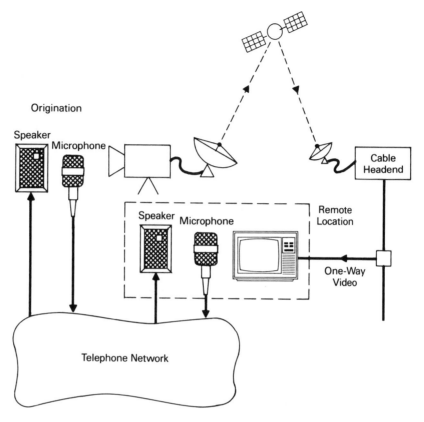

Fig. 4.11: Videoconferencing with one-way video, carried by satellite and cable, and two-way audio, carried by the telephone network.

Two-way audio allows viewers to talk back to the origination point, but they cannot be seen there. A cable system can serve as the last leg of this delivery system, carrying the video signal from a satellite earth station to one or several viewing locations. It makes possible immediate setup of a viewing location wherever there is cable access, instead of requiring the time and effort to set up a temporary satellite receiving station.

A more demanding form of business videoconferencing has users in different locations sending video, graphics, and audio in both directions. This requires an upstream as well as a downstream video channel in each of the cable systems. Cable systems are not implementing upstream video capabilities, although it is not impossible to do. Digital "compression" techniques can reduce the capacity requirement of a digitized television signal to as low as (or even lower than) 1.5 Mbits/second, a data rate which is already offered for private channels on a number of cable systems.

Educational and Social Services

Municipalities have mandated institutional cables to link together government offices, social services, schools, and medical institutions in addition to the standard residential cables. The institutional cable does not necessarily run along the same routes as the residential cable, and it is normally frequency split at a middle frequency, unlike the residential cable's low-end split, to allow a more balanced communications traffic in both directions.

Many institutional uses can be imagined, although few are yet implemented. Libraries can consolidate bibliographic and circulation functions, and even provide remote reading services for their patrons. Hospitals can exchange facsimile documents, X-ray images, and insurance forms, and perhaps televised operations and lectures as well. Schools can provide electronic learning in many forms, including televised classes, instructional software, and access to information resources. Government offices can coordinate activities, replace slow and expensive paper processing with electronic processing, and provide easier access for individuals and institutions utilizing government services.

Although many of the possibilities suggested here are from one institutional location to another, the institutional cable can be linked, within the cable headend or from hub to hub, with the residential cable and other communication systems, so that individuals in their homes can avail themselves of many of the institutional services. Special services can be provided for the elderly, disabled, and other population groups with special needs. Medical alert services, already being provided in elementary form on the residential cables of several interactive cable systems, could be expanded to support frequent monitoring of outpatient health. The potential services are not only for the ill or disadvantaged; professionals might enjoy services helping them to maintain proficiency and professional contacts.

Communication services will never replace face-to-face human contacts where they are really needed. However, a large amount of daily life, and particularly encounters with educational, medical, social welfare, and government services, involves undesirable travel and clerical effort. The potential of cable in this area remains to be realized, but is as large as that of any of the ancillary services.

The Internetworking Future

The examples in this chapter have a common thread of "interconnection," and strongly suggest that cable distribution systems have something to gain by becoming integrated into a much larger communications infrastructure. It is nearly impossible to keep any communication system isolated and local, even a one-way cable network. Users accustomed to the universality of the telephone network will not be forever satisfied with cable services which are constrained to a small geographic area or limited set of communication partners.

We can already see the outlines of the future integrated communications environment of the residential user, suggested in Fig. 4.12. An "intelli-

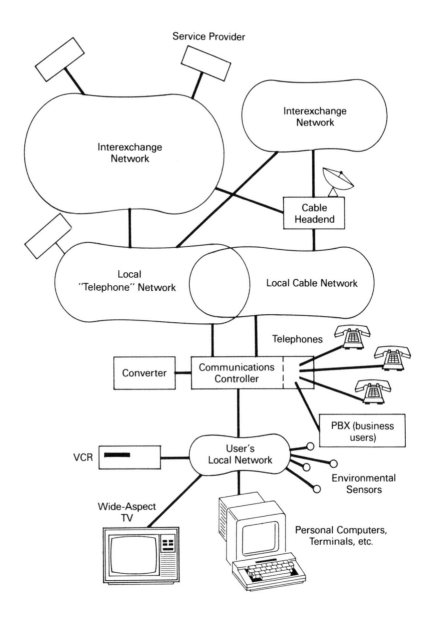

Fig. 4.12: *A guess at the future integrated communications environment, in which local cable facilities will be an invisible part of a mix of communications facilities serving a broad range of communications needs.*

gent'' communications controller will make connections with cable, telephone network, and possibly roof-mounted satellite receiver facilities. When this happens, the cable may well be optical fiber installed by the telephone network (see Chapter 12), or perhaps by a financial consortium which will lease capacity to both telephone and cable companies. Connections will be available for telephones, data equipment such as personal computers, video equipment, security systems, environmental management systems, utility meters, and whatever else is desired in local equipment which must communicate with the outside world. The controller will select the best and most economical communications medium for each function and make sure everything works properly together.

In this future scenario, which could become real around the turn of the century, cable will no longer be separated physically, although video entertainment may remain a distinct service. Communications, however, will be invisible to the user and transparent to the user's applications. Coaxial cable, having been installed at great cost, may persist into this future, but users will rarely, if ever, think about it as they move easily among entertainment, information, and personal communication services in an integrated communications environment.

For Further Reading

[1] *CableVision*, Apr. 18, 1983.
[2] "Two-way cable falters," *The New York Times*, Mar. 28, 1984.
[3] "Poor reception: Warner curtails QUBE," *Time Magazine*, Jan. 30, 1984.
[4] *The New York Times*, Mar. 4, 1984.
[5] R.W. Lucky, "GameNet," *IEEE Commun. Mag.*, vol. 15, pp. 14–19, Nov. 1979.
[6] "Two-way cable television services," LINK Resources Corp., New York, NY, Apr. 1984.

5 The Public Interest

The cable industry has grown up in a swirl of controversy over its development and operations, with concerns ranging from impacts on society and human rights down to the narrowest economic conflict. Issues such as subscriber privacy, control of program content, "must carry" rules, local versus federal regulatory authority, franchise awards, franchise fees and renewals, public and commercial access to channels, pole attachment fees, copyright fees, and competition with telephone companies continue to arouse intense feelings and generate Federal Communications Commission (FCC), judicial, and legislative proceedings.

Some of these issues are examined here directly, and some through a discussion of FCC rules and federal legislation. The federal government has been erratic in its policy toward the industry, but on the whole has been moving toward deregulation, as exemplified by favorable court rulings, the FCC's gradual lifting of most regulatory constraints and the emergence from Congress of the historic Cable Communications Policy Act of 1984, or "Cable Act." The industry itself has had mixed feelings about regulation, opposing restrictions on what cable systems may do with their facilities and charge their customers, but favoring franchise exclusivity and renewal standards, controls on competitive private cable operations, and a nation-wide standard.

The cable industry's stand against restrictive regulation was fortified in mid-1984 by a landmark decision of the U.S. Supreme Court forbidding state or local regulation of the content of satellite-delivered programming and by an FCC ruling removing all such programming from the "basic" tier of programs subject to local pricing regulation. As an NCTA spokesman said at the time, "Operators won't put their heads back in a regulatory noose that the Court said we don't need" [14] (see "For Further Reading" for numbered reference). An even larger victory came in July 1985 when a Federal Appeals Court struck down the FCC's "must carry" rules, which had required cable operators to carry all local broadcast stations, occupying channels which might otherwise be used for more salable satellite-delivered programming. The Cable Act, described later in this chapter, lifted most local regulatory control, except for services outside of traditional video entertainment and some relatively limited constraints in other areas. The Cable Act was intended to end a period of bickering and uncertainty in the relations between cable systems and local regulatory authorities, but many of its provisions are less than crystal clear. It may not stop continuing disputes, in the courts and before the FCC, on franchising requirements and other controversies.

Broadly speaking, the public interest is that cable communications enhance our personal, community, and professional lives. In the early years, when cable seemed to be something out of a science fiction future, social scientists and others were sometimes ecstatic about the opportunities for enlightenment and participatory democracy, particularly in interactive cable television, and sometimes fearful of the demagogic, repressive, and divisive possibilities which are the other side of the coin. As Ithiel de Sola Pool, the late M.I.T. authority on communications policy, wrote in 1973,

> ...there is no distinction between the electronics that make demagogy easier and those that make responsible politics easier. It is the men who use them that make the difference. Modern society needs better modes of communications between the people and those in power. It needs to lessen the citizen's sense of alienation, his sense of powerlessness and isolation. To do so it must provide communications that are faster, more individualized and more responsive than are the mass media today. Electronic, computer-controlled, broadband, two-way facilities can make such communications possible. But whether the outcome of that new technology will be the "dark night" of ideological manipulation or the responsible politics of liberal discourse depends not on the technology but on men. [1]

The most pessimistic scenario was that of "big brother" thought control through interactive cable, as suggested by George Orwell in his book *1984*:

> With the development of television, and the technical advance which made it possible to receive and transmit simultaneously on the same instrument, private life came to an end. Every citizen, or at least every citizen important enough to be worth watching, could be kept for twenty-four hours a day under the eyes of the police and in the sound of official propaganda... [2]

So far, cable has come nowhere near realizing its potentials for either enlightenment or repression. One of its problems is, in fact, that it has not yet been able to realize any of its potentials, even for entertainment, where its image is about the same as that of ordinary television. Perhaps it is still too early. As newspaper columnist Ron Legro wrote in 1984, in regard to public access,

> The masses had to be taught to take advantage of libraries; thereafter, libraries became as essential to education (and by extension to society) as blackboards. Public access channels and other "profitless" features of cable have even greater potential—after all, the cable system could transmit entire libraries of data. But this potential will remain unrealized until it is available to most citizens. [15]

But some progress has been made, and John Fjiorescu, a media adviser to political candidates, was able to claim, in early 1984, that

> Cable democratizes the use of television. Until now an appearance on TV was a privilege reserved for multimillion-dollar campaigns. Cable enfranchises a new class of office-seeker, right on down to the water commissioner. [18]

Editorial Control

Although the "democracy-tyranny" debate has fizzled out, fierce battles are still being fought over bedrock constitutional questions. One of them is the question of editorial control, pitting rights of a cable operator to control channels and program content, just as an editor controls what is in a printed publication, against demands for public access, political fairness, different programming, and the elimination of "indecent" programming. The argument for constraining an operator's editorial control is that a cable system, unlike a newspaper, is a unique public utility, an assertion which cable operators strongly deny.

Sometimes the technical realities of cable broadcasting influence government policies. The FCC, for example, in considering extension of broadcasting's "Fairness Doctrine" to cable systems, saw tremendous difficulties in administration. As William Johnson, Deputy Chief of the FCC's Mass Media Bureau, expressed it,

> The cable operator, as a practical matter, has no control over any of his stuff. He would have to sit there with 50 screens in front of him watching all the programming to make sure they are fair...As soon as we're saddled with this, it's going to be unbelievable. [18]

Although the issue of editorial control is far from resolved, some important decisions have been made. The U.S. Supreme Court, in the Capital Cities versus Richard Crisp case referred to earlier, ruled unanimously in June 1984 that states could not require operators to alter the satellite-distributed programming they obtained from national networks. The case concerned an attempt by the State of Oklahoma to force cable operators to edit liquor and wine commercials out from such programs, something virtually impossible to do. Justice William Brennan said that if cable operators had to totally abandon the carriage of distant signals to meet the Oklahoma prohibition, "the Public may well be deprived of the wide variety of programming options that cable systems make possible" [16]. This ruling was viewed favorably by cable operators, and was taken by some to imply judicial disapproval of content restrictions of any kind.

A related content issue is that of cable carriage of entire programs which some subscribers do not like. Many communities attempt to impose local standards of "decency" on their cable systems, and the claim is often made that CATV should be subject to the same restrictions as those imposed on broadcast television. Cable operators answer, as in the following comment of NCTA attorney Robert Roper:

> The rules that apply to broadcasting have no relevance to cable, because the public can pick and choose. Cable isn't burdened by the broadcaster's inability to protect the unwilling child or adult from unsuitable programming...cable has the technology to insulate you from services you don't want. [17]

Programs with a sexual orientation are economically important to some cable operators and anathema to some citizens, legislators, and journalists, including the columnist Jack Anderson, who wrote that the Supreme Court's Oklahoma decision was a

> blessing to home pornography...the way the decision was worded, it would bar states or communities from regulating the content of cable television programs. This may give the green light to a few depraved profiteers, who use cable TV to pipe the most disgusting sex scenes imaginable into American homes...sex orgies that would make the Caesars blush are shown coast-to-coast on cable TV. [19]

Although most challenges to program content are local and judicial, the State of California has considered legislation which would establish some level of restriction, ranging up to making cable systems directly responsible to their communities for programming decisions. Efforts such as these, if they should ever succeed, would probably be found unconstitutional, provided cable operators can make a convincing argument that subscribers are given adequate means, through parental control devices (now mandated by the Cable Act), for program control in their homes and are protected against unintentional leakage of undesired programs into their viewing channels. Video or sound leakage from a supposedly excluded channel is a technical problem in many systems.

Channel *access* is another area of constitutional conflict. It concerns the cable operators' obligations to provide studio facilities and cable channels for public use, and lease channels to commercial programmers and service providers. Facilities for public, educational, and government access were once mandated by the FCC and are still demanded by most communities, although cable operators building the expensive new systems have been anxious to cut back on their initial commitments in this as well as other areas. But the view that larger, many-channeled cable systems should give up some of their channels to others has prevailed. Provision of a modest number of leased channels was, despite the opposition of the cable industry, written into the Cable Act, and franchising authorities are free to demand public, educational, and government access as a condition for granting a franchise.

Privacy

Personal privacy is another major "rights" issue for CATV. Privacy means different things to different people, but in general it refers to the right to keep one's person, vital data, and life history away from the eyes and out of the hands of unauthorized persons or institutions. Sometimes one is willing to give up some privacy in order to obtain a benefit, such as financial credit, but this is an authorized and selective concession rather than an unauthorized invasion. Some of the conceivable invasions of privacy on cable are:

1) A cable operator could acquire an economic and psychological profile of a subscribing household which could be used to support sales or other efforts focused on that household.

2) Information on a subscriber's selections of pay programming or other services could be used against him politically or professionally.

3) Funds transfer data, votes, and other private traffic originating from a subscriber on an interactive system could be intercepted by unauthorized persons or misused by authorities; and

4) The subscriber's activities and movements could be monitored by "big brother" listening, sensing, and transactional devices connected through the cable.

Although there has been no compromise of the individual privacy of cable subscribers serious enough to receive national attention, the potential is there and is of concern to legislators, regulators, the cable industry and the public. The Cable Act specifies that

A cable operator shall not use the system to collect personally identifiable information concerning any subscriber without the prior written or electronic consent of the subscriber concerned,

and has stringent limitations on disclosure of information and notification to subscribers of what is being done with their information. Some cable operators, most notably Warner Amex with its early Code of Privacy, took their own steps to control use of personal information within their systems.

The Big Brother potential of cable was also in the minds of the legislators when the disclosure provisions were being developed. In consideration of the privacy issue in the House of Representatives, lawyers for the ACLU

successfully lobbied the House Commerce Committee to adopt a provision sharply limiting the right of law enforcement agencies to obtain information collected about subscribers by cable television systems...law enforcement agencies could obtain a warrant for personal information controlled by a cable company only after showing clear evidence that the subject was involved in a crime and that the information they were seeking was directly related to their investigation. [22]

The final wording in the Cable Act, much of which is given later in this chapter, went a long way toward visualizing and heading off the possibilities for invasion of subscriber privacy on cable.

Security of cable communications is a related but not identical concern covering reliability questions as well as protection of facilities and information against intentional attacks. The security susceptibilities include:

1) Malfunction of an alarm or environmental management system, leading to damage to a subscriber's property. An example would be turning off a heating plant on a cold winter's day.

2) Garbled transmission of transactional data, leading to incorrect billing or other error.

3) Inadequate protection of customer files from either unauthorized access or accidental loss or damage.

Good engineering and management practices, which have not always been characteristic of cable systems, and sophisticated data encoding techniques may possibly reduce these dangers to an acceptable minimum. But cable systems have not usually been built to the rigorous standards of commercial communication systems. If interactive services develop, regulatory authorities may find it necessary to issue new rules and regulations in this area.

Economic Disputes

The most hotly disputed public interest issues are, as always, economic. In general, cable operators favor a fair competitive environment, but that is not always easy to define. The cable industry wants a maximum of commercial freedom for itself but is not so sure that regulatory constraints should be removed from its major communications competitors, the broadcast, private cable, and telephone industries. It usually prefers exclusive franchising, although many operators would give up exclusivity in return for full First Amendment freedoms from municipal regulation, but when it has exclusivity, it does not want to be classified as a public utility or "natural monopoly." It wants limits on taxes imposed by local authorities, and on the rates charged by telephone and power companies for use of their poles for stringing its cables, but few or no constraints on the rates it can charge for its services. It feels it should have access to potential subscribers in apartment buildings despite the alternative arrangements, often involving personal income, which building owners may make with private cable (SMATV) operators. It wants to pay minimal copyright fees and enjoy a favorable split of pay revenues with the actors, producers, and primary distributors of video entertainment. Many of these are, of course, negotiating demands which have already been compromised in the legislation and regulatory outcomes of recent years or will be worked out in the marketplace.

The dispute over ancillary data transmission services is one of the more significant and clear-cut of the inter-industry conflicts. A number of cable operators offer leased point-to-point data circuits to commercial users, similar to those long offered, under regulated tariffs, by telephone companies. Some cable operators, especially Cox Cable, have emphasized "bypass" transport from the offices of large customers to the networks of long-distance communications companies, as described in the last chapter.

The telephone companies, understandably upset by this previously unregulated competition to their bread and butter businesses, pressed hard to subject CATV data and voice services to the same state regulation exercised over intrastate communications provided by telephone companies. The telephone industry's expressed fear, as stated in 1983, was

> At stake, in the view of AT&T and the National Association of Regulatory Commissioners, is the prospect of cable companies taking enough revenues from local phone companies to prompt further local phone rate increases. [20]

Estimates of the potential revenue loss have been hard to come by, but Pacific Telephone officials estimated in 1984 that $760 million in revenue would be at risk if unregulated competition from CATV operators were permitted.

State public utility commissions have usually sided with the telephone companies. In a case in Omaha, Nebraska, the Nebraska Public Service Commission ruled that Cox Cable of Omaha, in supplying voice and data services through its CommLine and Indax systems, was a common carrier subject to tariff regulation by the Commission and must cease and desist from such operations. The cease and desist order was reversed by a federal district court, but the judge, refusing to rule on the question of whether or not cable data communications is subject to regulation, returned the case to the Commission for further consideration.

The question of state versus federal jurisdiction was, for this question, apparently resolved in the Cable Act. Although the attitude of Congress was at first sympathetic to the cable industry's point of view, exemplified by the comment of Chairman Timothy Wirth of the House Telecommunications Subcommittee that

> If cable companies choose to enter the market, their presence in no way threatens the present monopoly of local telephone companies or universal service [21],

the final legislation came out much closer to the telephone industry's position. A compromise was worked out among the Pacific Telesis Group (parent of Pacific Bell Telephone), the NCTA, and the Communications Workers of America in the summer of 1984. Under the compromise, the only clearly unregulated data transmission services to be allowed on cable would be those related to provision of video programming, voter preference polls, video rating services, and one-way teletext and two-way videotex services such as a weather channel, stock market information, and airline guides and catalogs. Electronic mail, one-way and two-way transmission of data and information not offered to all subscribers, home banking and shopping, data processing, videoconferencing, and voice communications would all be subject to state regulation at the option of state public utility commissions. Even if a state PUC were not to exercise this option, cable operators providing data services would have to file informational tariffs so that their competitors would be aware of their activities.

Although a state may regulate commercial communications services on cable, the FCC ruled in August 1985 that it may not prevent a cable operator from entering the market. The FCC preempted Nebraska's regulation barring Cox Cable from offering intrastate high-speed voice and data services in competition with local telephone companies. The FCC may yet preempt state regulation entirely.

A second question which has stirred up many CATV-Telco (and CATV-Power Company) disputes is that of the fees paid by cable operators for the privilege of hanging cables from utility poles. The utilities naturally want to assess the highest possible fees, especially if the cable operator competes with some of the utility's businesses. The FCC has claimed jurisdiction when state regulators do not promulgate appropriate rules and regulations, and has drawn up a pole attachment fee formula, simply expressed as

$$\text{Maximum rate (\$/year)} = \frac{\text{Space occupied by CATV}}{\text{Total usable space}} \times \binom{\text{Yearly operating expenses}}{\text{and capital costs of poles}}$$

which, in the words of Congressman Wirth, "...is fair to all parties and has worked extremely well" [23]. Most of the dispute over the pole attachment formula has not been over the formula itself, but the definitions of its components. The FCC estimate of 13.5 feet of usable space per pole was contested in 1984 by Interstate Power Company (in Minnesota), whose $4/year/pole attachment fee was being challenged by Group W Cable. Interstate claimed that 10.24 feet of usable space per pole was a more accurate figure, but the FCC stood by its own estimate, and for this and other reasons reduced the attachment rate to $1.36. In another case in Florida, the FCC ordered a reduction from $5 to $2.46.

Pole attachment disputes have concerned access rights as well as rates, with service competition the real issue. In 1983, pole access was for a time denied to Cable Systems Pacific by Pacific Northwest Bell,

> taking the position one, that the provision of data transmission services is a regulated activity, and two, that the state public utilities commission has authority to regulate it. [12]

The telephone company made clear its belief that

> what they're offering under the rubric of institutional services is an exact carbon copy of what we offer, and that is private line data and voice services. They call them data services, but they can be made voice services by changing the modem on the other end of the line...it doesn't have anything to do with television; that's our concern. [12]

The response of the NCTA was that "no matter how creative their rhetoric or alleged justifications of their legislative position," the Pacific Northwest Bell position "makes clear the real goal of AT&T and the Bell Operating Companies is to shut down potential competition from cable operators through any means possible" [12].

The FCC's watchful eye on pole attachment fees has been characteristic of its attitude toward cable, which in recent years has leaned toward protecting cable and other new media from unfair economic burdens which might be imposed by utilities or municipalities. State utility commissions, although generally far less sympathetic to CATV interests, have supported CATV positions against private cable systems (see Chapter 9) and for exclusive franchises. The question of CATV access to residents of apartment buildings in which private cable systems have signed service contracts with landlords is a delicate one, with just compensation to landlords a large part of the problem. Franchise exclusivity is almost universally observed, with few overbuilds in existence and fewer still profitable to the two companies serving the same area. However, the Cable Act specifically permits award of more than one franchise in a given market, and a 1985 federal appeals court ruling which could end exclusive franchising in Los Angeles was seen as a victory by some industry spokesmen, who felt that it could also free cable systems from municipal franchise requirements.

Other economic issues of considerable importance to the cable industry and the public will be introduced in later parts of this chapter. Many of them are in the area of local regulation, and can be grouped under the heading of "franchising," which covers the competitions, agreements, regulatory and renewal policies, and fees associated with a community's grant of operating franchises to one or more cable operators.

Franchising

The areas of responsibility left to local and state authorities—franchise award and renewal, subscriber rates, data communications services, taxation, pole attachment, theft of service—have already been discussed to some extent. Franchise issues have always been among the most difficult and complex regulatory questions. The awarding of franchises has a long and somewhat unsavory history, and although the franchising era is largely past, there remains a legacy of tension and mistrust between cable operators and the municipalities in which they are licensed to operate.

Provided a cable system is to cross public property, the municipality in which all or part of the system is located has the power to grant an exclusive operational franchise, demand payment of fees, and negotiate whatever other terms and conditions it desires, such as basic service rates and public access facilities. The cable operator gains insulation from competition for a franchise period of typically 15 years.

This is not to say that there hasn't been controversy over the reach of municipal authority or the legality of franchise exclusivity. The private cable industry (see Chapter 6), which installs and operates Satellite Master Antenna Television (SMATV) systems in apartment buildings, claims total immunity from municipal regulation, including fees, if its facilities do not cross public property. The franchised cable industry regards private cable as an intolerable "cream skimming" operation. Private cable has not yet definitively won or lost, but the trend seems to be for it to remain unregulated while conceding to the franchised cable operator the right to solicit subscriptions from building residents.

Efforts to compel cities to award overlapping franchises to several cable operators, that is, to do away with exclusivity and permit overbuilds, have been less successful. The Mountain States Legal Foundation in 1982 filed a lawsuit in a federal district court challenging Denver's award of an exclusive cable franchise to Mile Hi Cablevision. The Foundation's position, according to Chairman David Flitner, was that

> with government and large cable companies acting in tandem, the free market has no chance to work its magic. Potential competitors seeking market entry and subscribers deprived of a choice are left out in the cold...the restoration of free and open competition in cable will come only from a comprehensive, constitutionally based challenge to the very authority on which franchises and regulation are based. [8]

The suit failed, with a 1983 decision supporting the right of local officials to decide for a "natural monopoly" if the economics of system construction point in that direction, as they usually appear to do.

A single franchise does not always cover the entire area of a single municipality. Fig. 5.1 illustrates the franchise boundaries of the four operators authorized by the city of Houston to operate there. New York City, after years in which only Manhattan had cable service, signed agreements in 1983 with six cable companies for eight franchises in the remaining four boroughs. The companies are expected to spend $1 billion to develop the new franchises over a period of seven to eight years.

Leaving aside legal challenges to exclusivity, the franchising process would appear to be straightforward and uncontroversial, as with any project for which competitive bids are requested. Instead, it has been a source of rancor and friction between cable operators and city governments, a prime cause of the later economic difficulties of some of the larger cable operators, and the focal point of allegedly unethical business practices in the industry. The

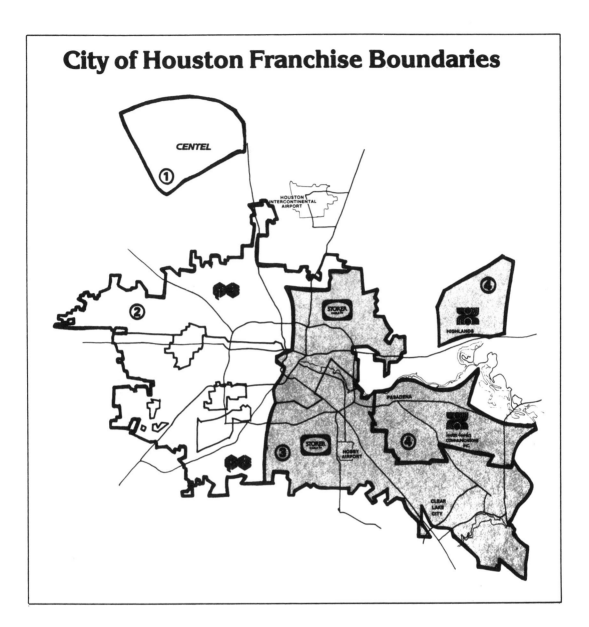

City of Houston Franchise Boundaries

Fig. 5.1: *Franchise boundaries in the city of Houston. (numbers 1 through 4). Source: CableVision, June 20, 1983 (with permission).*

large, undeveloped urban areas looked so attractive to cable operators in the 1970's and early 1980's that they were willing to go to almost any extreme to win franchises. City governments and officials, sensing this greed, developed some of their own. The situation was well described in a study sponsored by the National Telecommunications and Information Agency:

> Important problems have arisen in the current franchising scene in major cities and metropolitan areas...Some of these problems are

associated with ever-escalating cascades of demands for services on the part of the cities and ever-escalating promises of services on the part of cable companies. Others arise from the intensity of the competition for these franchises. This has given rise to a series of questionable practices whose extent and pervasiveness is perhaps without precedent in the dealings of a national industry with local governments. [9]

One questionable practice which received a lot of media attention was called "rent-a-citizen." This meant enlisting the aid of influential local citizens or institutions in a franchise competition by giving them a substantial equity interest in the proposed cable system. The "rented" citizens were expected to exert their influence on their friends in municipal government. "When you're trying to win the hearts and minds of 10 city councilmen," said Gustave Hauser, former chairman and CEO of Warner Amex Cable Communications, "you obviously want someone who knows them, looks good and so forth" [25].

Sometimes these local citizens were asked to make a token investment. Warren Buffett, chairman of the Sun Newspapers in Omaha, was one of those who declined:

> In a three-page letter to the publisher of a newspaper of which I am board chairman, one of the major national cable operators (and one of the six final applicants) asked us, either individually or corporately, to join up with them for a significant piece of the action. Our financial commitment would have required us to forego a night out at MacDonald's. [10]

Buffett analyzed the financial particulars of the eventual award of the Omaha franchise to a subsidiary of Cox Broadcasting (not the company that approached him). Eight local citizens, for an aggregate investment of $200, obtained a 20 percent interest in a franchise requiring a $37 million investment for its construction. This 20 percent interest was to have, by Cox's projections, a value of about $12 million ten years after award of the franchise. Warner Amex, perhaps more discreet than other operators, gave away 20 percent of the equity in its former Pittsburgh franchise to 17 local minority groups rather than to individuals. The FCC was remarkably unconcerned. "The Vanderbilts were selling railroads at the turn of the century and they bought state legislatures," said Willard R. Nichols, chief of the FCC's cable television bureau. "Now they (the cable companies) are buying city councils. It's getting a lot of press, but I'm not sure it's much different from the way it's always been" [25].

Aside from the initial improprieties, citizen-renting damaged the financial prospects of the franchise-winning cable operators and raised the possibility of higher costs to consumers. That era passed some time ago, but an even greater threat to the economic viability of the industry continued in the demands of cities for high fees and extensive services and in the willingness of some cable operators to promise anything. In the early 1980's competitions, Boston wanted an 8 percent fee (on gross revenues) and a basic subscription fee of only $2. Many cities demanded several public-access studios with full color television capabilities, and basic service including a substantial number of public-access channels. Others wanted job training programs and other special "sweeteners."

Some of the MSO's finally decided that the franchise competition was not worth it, and pulled out. "We looked at the crazy demands of the cities, the political opportunism, and the eagerness of cable operators to win at almost any price," said ATC's president, Trygve Myhren, in 1983, "and decided there was no way to make the risks pay off" [11]. Others who stayed in too long and must now pay the price are admitting their mistakes publically. Drew Lewis, who joined Warner Amex as Chairman after its large franchise commitments were made, said "I'll tell you what has happened to us and to everybody else who has obtained major city franchises. We've promised too much" [3]. Perhaps this era, too, is now over.

The conflict between cities and cable systems over franchise rates, franchise renewal, and city control over subscription rates and operating conditions may have been ameliorated by the 1984 Cable Act, but it will never completely disappear. There will be differences in interpretation of the new law, and continuing pressure on cities from private cable, regional business communication networks, telephone companies, and other cable competitors to be less sympathetic to cable interests. Some of the few municipalities not yet committed to a franchised operator may investigate municipal ownership possibilities, as Belmont, Massachusetts, did in 1983. But as cities and franchised cable operators more and more see a mutual interest challenged by alternative distribution media, they may come together as allies in the Communication Wars of the future.

FCC Rules

Although some of the important FCC cable rules were implicitly introduced in the last section, most were not. All of these rules are, of course, intended to be in the public interest. Although, with the Cable Act, FCC rules may not continue to have the large impact on the development of cable technology and the cable industry's relationships with broadcasters, program producers, and the general public which they have had in the past, the discretionary powers left to the FCC are still very large.

Most of the rules fall into the following major categories:

1) Division of Regulatory Responsibility Between FCC and Local or State Authorities

The FCC is given jurisdiction over technical quality, compliance with affirmative action law, and a number of other questions concerning signal transmission and the relationships between CATV and television broadcasting. Regulation of franchising, including franchise fees, public access channels, pole attachment and, until January 1987, basic subscriber rates, are left to local and state authorities.

Some of these jurisdictions were different before passage of the Cable Act. Between 1972 and 1977, the FCC set franchising standards, including construction timetables, procedures for franchise awards and renewals, and basic subscriber rates. Franchise fees were "capped" by the FCC at 3 percent of gross revenue under normal conditions, and 5 percent, by special authorization, if the higher rate was required to cover the cost of local regulation. A flat 5 percent cap is specified in the Cable Act. Non-mandatory franchising standards, including suggestion of a 15-year franchise period, are still found in the Rules.

2) Business Operations

The FCC rules have long provided for registration, record keeping, annual reporting, equal opportunity hiring, and adjudication of complaints. The registration and reporting requirements have been reduced and simplified in recent years, with financial reporting eliminated entirely in 1983. The only reporting specifically mandated in the Cable Act is of employment information for evaluation of equal employment opportunity compliance, but licensing and other requirements call for additional data.

3) Carriage Rules

The FCC, seeking to protect local television broadcasters in its 1972 cable rules, forbade cable systems from "importing" and retransmitting more than one distant broadcast signal of a full network station of each national television network. One to three "independent" stations could be imported, depending on the size of the market. These import restrictions were dropped in 1981 because of the benefits signal importation could bring to consumers "without harming the ability of broadcast stations to meet their public interest responsibilities."

Cable operators are, however, still required to provide non-duplication protection to local television stations within whose service zones a cable system operates. That is, the cable system must blank out duplicative programming on imported signals, except for imported signals which are "significantly viewed" in the community. The cable system must also do its part to maintain local television sports blackouts by not carrying local sports events shown on imported signals when they are not shown on local television broadcasts.

An earlier additional restriction on non-rebroadcasting of programming syndicated to the imported station ("syndicated program exclusivity") was dropped in 1981. Obscenity, to be defined by community standards, remains prohibited, now under terms of the Cable Act, on basic channels, but sexually explicit programming for pay subscribers is not prohibited on the federal level if technical safeguards against unintended viewing are provided. *Local* lawsuits against cable operators for carrying allegedly obscene movies have occurred. These cases are similar to those involving theater operators, and theaters and cable operators under attack frequently cite each other's activities as examples of community acceptance of sexually explicit programming.

On the other side of the carriage rules were the controversial "must carry" requirements, overturned in July 1985 by the U.S. Court of Appeals in Washington. These were intended for protection of local broadcast television stations, which a television set connected to a cable cannot directly receive unless some means is provided for switching to a receiving antenna. They mandated cable carriage, on broadcaster request, of

- any TV station licensed to communities within 35 miles of the cable system community;
- noncommercial educational TV stations within whose secondary ("grade B") reception contour the cable system is located;
- most commercial and noncommercial local translator (rebroadcaster) stations;
- any station whose signal is "significantly viewed in the community";
- certain additional grade B signals for systems in smaller markets.

Cable operators deeply resented the "must carry" rules. As Gustave Hauser put it

Cable programmers are currently penalized by a second-class status imposed by the FCC's "must carry" rules, which require a cable system ... to carry even the least watched or least meritorious broadcast signal ahead of the best cable-satellite signal. Where channel capacity is limited, a cable operator must devote his precious channels to broadcast programming without regard to the desires of his customers. This regulatory preference is both arbitrary and a denial of equal protection of the law to cable programmers. [6]

The demise of the must-carry rules came in a case brought by Quincy (Massachusetts) Cable TV, Inc., and Turner Broadcasting. The U.S. Court of Appeals ruled that must-carry violated the First Amendment rights of operators and programmers. The National Association of Broadcasters was extremely disappointed with the ruling, and indicated it would fight the decision and press, in Congress, for elimination of the compulsory copyright license described later in this chapter. If the compulsory license were eliminated, cable operators would pay individually negotiated market rates for imported broadcast programming instead of the flat fees associated with the compulsory license.

Additional carriage rules specify that cable systems are not required to carry scrambled subscription television (STV) signals, and have no restrictions on carriage of radio stations. Previous requirements for public access channels were dropped in 1981, although public, educational, and government access can be, and usually will be, negotiated in franchise agreements between cable operators and local franchising authorities. Requirements for leased access channels were established in the Cable Act and are described later.

4) Two-Way Capability

A Supreme Court ruling in 1979 did away with the FCC's two-way capability rules, giving that regulatory authority to state and local agencies, but the FCC strongly supported the development of interactive services in the prior decade. It promulgated a rule in 1972 which required two-way capability on all new CATV systems of a minimum size. The FCC's view was cogently summarized in a letter sent the year before by Dean Burch, FCC Chairman, to Senator John Pastore, Chairman of the Senate Subcommittee on Communications:

> After studying the comments received and our own engineering estimates, we have decided to require that there be built into cable systems the capacity for two-way communication. This is apparently now feasible at a not inordinate additional cost, and its availability is essential for many of cable's Public Services. Such two-way communication, even if rudimentary in nature, can be useful in a host of ways— for surveys, marketing services, burglar alarm devices, educational feedback, to name a few. Of course, viewers should also have a capability enabling them to choose whether or not the feedback is activated. [1]

Many of the cable systems built subsequent to the FCC ruling did not, in fact, support interactive services. Trunk amplifiers might come in housings which could accommodate upstream amplifiers, but the upstream amplifiers were not necessarily there. However, the provision of interactive communications has been an important element of the major franchise competitions of the early 1980's. Although not many systems actually have it, some form of interactivity is likely to become pervasive as per-per-view services develop. Regulatory authorities may soon be faced with the problem of cross-media regulation of interactive communications realized in a hybrid CATV-telephone network system, as described in Chapter 2.

5) Pay Cable Regulation

Pay cable subscription rates have been deregulated for some time. Pay cable program *content* was removed from FCC jurisdiction by a federal court decision in 1977.

6) Ownership

Although the FCC's previous restrictions on cross ownership by television station operators were dropped in 1982, the Cable Act imposed such a restriction, as well as one on cross ownership with common carriers, e.g., telephone companies, except under conditions described later. There are no restrictions on telephone company provision of cable facilities to cable operators, on a leased or tariffed basis. Telephone companies may build and own cable systems but not operate them, except in some rural areas, as specified in the Cable Act.

7) Pole Attachment

State public utility commissions, previously required to certify to the FCC their authority to regulate pole attachment rates (charged by telephone or power companies), have what appears to be a similar requirement under the Cable Act to issue appropriate rules and regulations or lose the right to regulate. In the past, when the FCC stepped in, it used the rate formula, given earlier, which assigns costs on the basis of relative occupancy of space on a utility pole. The FCC formula was approved by Congress in the Pole Attachment Act of 1982, and does not appear to be challenged by the new law.

8) Technical Requirements

Aside from requirements on the technical quality of retransmitted broadcast signals, the FCC maintains stringent standards on signal leakage, and must be asked for special permission for use of frequencies in the 108–136 MHz and 225–400 MHz bands, used by aeronautical and emergency services. Several specific emergency frequencies are completely off bounds. Cable Television Relay Service (CARS) stations, which are microwave relay stations used to bring in programming to headends, must be licensed by the FCC. The frequency band available for such relay stations was expanded in 1979 to the current 12.7–13.2 GHz.

9) Copyright

The 1976 federal Copyright Act [7]

established for the first time an obligation for cable operators to pay copyright fees for the retransmission of broadcast signals. [5]

The law provided for a compromise royalty system which freed cable operators from having to obtain a release each time a program is retransmitted. This compromise is the ''compulsory license'' referred to earlier, which in effect compels program owners to allow cable systems to carry their programs. Under this license, a cable operator pays a flat fee twice a year into a fund maintained by the Copyright Office of the Library of Congress, and distributed by the Copyright Royalty Tribunal, an entity created by the Copyright Act. Cable operators contributed $80 million to this fund in 1983. No fees are paid for rebroadcast of local signals, which had to be carried until the summer of 1985. Thus the compulsory license although strictly speaking outside of FCC jurisdiction, was closely coupled with the must-carry rules.

Although content with the compulsory license arrangement, the cable industry was not happy with the Copyright Royalty Tribunal, which could increase rates. Broadcasters, as already noted, did not want to keep the compulsory license arrangement without must-carry rules for local signals. The result was activity in Congress to change the copyright law once more, although no new legislation had appeared when this book went to press.

Cable copyright fees were fixed by the Copyright Act in a formula depending on a cable system's gross revenue from the retransmission of broadcast (not pay) signals and on the number of "distant signal equivalents," i.e., non-network distant television stations imported by the system. The operator originally had the choice of paying 0.675 percent of gross revenue (as defined above) for the privilege of carrying any number of distant signal equivalents, or paying on a sliding scale, from 0.675 percent down to 0.2 percent, for each distant signal individually. The Copyright Royalty Tribunal changed the formula several times, so that, by May 1985, the sliding scale ranged from 0.893 percent down to 0.265 percent. Larger systems could carry up to three distant signals without paying any fees, and pay a 3.75 percent fee on revenues attributable to carriage of additional distant signals beyond the first three. The smallest systems were allowed to pay a semi-annual flat fee of $28.

The Cable Communications Policy Act of 1984

The effort made in 1983–84 to pass national cable legislation succeeded in the last days of the 98th U.S. Congress. The new law attempted to put an end to many years of contention between cable operators and the public authorities with which they must coexist, and laid the foundations for an era of stability and orderly development for the cable industry. It also established certain rights for subscribers, although they could find themselves paying more for basic cable services as prices come to more nearly reflect costs. The Cable Act was an important amendment to the Communications Act of 1934, an old statute which still governs most communications activity. The passage of the Cable Act did not, however, overturn the FCC jurisdictions and specific rules enumerated earlier. Almost all of the pre-Cable Act FCC rules continue in force, as do many state and local regulatory powers, but the regulatory authority of each level of government was greatly clarified.

The main thrusts of the new law are:

- Local franchising authorities continue to regulate cable systems through the franchise process, with federal standards for awarding franchises and a prohibition on excluding an area because of the income of its residents. In a request for proposal for a franchise or franchise renewal, requirements may be set for facilities and equipment, but not for offering video programming or other information services. Provision for enforcement of construction requirements, delivery of video programming (once offered), and customer service agreements may be required by franchising authorities.
- The FCC continues to have rulemaking authority over technical standards and most other matters which it was regulating prior to passage of the Act.
- Franchising agreements may, but do not necessarily have to, require the designation of channel capacity for public, educational, and government use. All existing requirements for public, educational, and government access remain in force.
- Local franchising authorities can charge cable operators franchise fees (taxes) of up to 5 percent of annual gross operating revenues; fees can be used by the franchising authorities for any purposes, and increases in franchise fees can be passed through to subscribers and separately billed. The FCC is denied any authority in this area.
- Subscriber charges for basic cable service are to be deregulated (removed from local franchising authority control) after two years, i.e., in January 1987, except in markets for which the FCC determines that there is insufficient competition. This is a highly controversial question, since the number of markets which remain regulated depends on the severity of the effective competition requirement. The presence of four television signals of at least grade B strength (secondary reception area), including affiliates of the three major broadcasting networks, was the FCC's first proposal for a definition of effective competition, but the NCTA argued that the test is too rigid, and the National League of Cities that it is not rigid enough. For markets with insufficient competition, the FCC is to prescribe regulations allowing local franchising authorities to continue regulation of basic services.

During the two-year transition period ending January 1, 1987, franchising authorities can continue to control charges for installation and rental of subscriber equipment and for basic cable service, "including multiple tiers of basic cable services," except that basic cable service rates, if not specifically frozen in an existing franchise agreement, can be increased by up to 5 percent per year without the approval of the local authorities. All retiering or repricing of cable services made in 1984 following the FCC's ruling permitting removal of satellite-delivered programming from the regulated basic tier are allowed to stand, as are all state deregulation statutes. The FCC is to submit a report to Congress on rate regulation within six years of enactment, with recommendations, if any, for new legislation.

• A cable company may obtain, from local authorities, modification of the terms of its franchise agreement on services, facilities, and equipment, including provision of public, educational, and government access facilities and equipment, if those terms prove to be commercially impractical. A public proceeding is to be held to consider such a request, with the issue going to court if the request is denied. The court "shall grant such modification only if the cable operator demonstrates...that the mix, quality and level of services required by the franchise at the time it was granted will be maintained after such modification." A cable operator may unilaterally replace or drop a particular cable programming service required by the franchise agreement in circumstances beyond the operator's control, including substantial increases in royalties to program suppliers.

• Franchise renewal, assuming no quick agreement between the franchising authority and the cable operator, is to be based on an administrative proceeding in which a cable operator's past performance and new proposal are reviewed. Renewal must be granted unless the operator has not substantially complied with existing franchise terms, is no longer financially capable, or has presented an unreasonable new proposal. Renewal may not be denied on the basis of noncompliance unless the franchising authority has given the operator notice of a violation and an opportunity to end the violation. Evidence from the administrative proceeding is admissible in court if there is still no agreement. If a franchise renewal is ultimately denied, the transfer of ownership of the cable system shall be at fair market value. The sense of the renewal provision is that a cable operator is to be protected from an unfair denial of franchise renewal.

• Leased channels are to be made available in larger systems. Cable systems with 36 to 54 activated channels must make available for lease 10 percent of the channels not dedicated to mandated use, and systems with 55 or more channels must make available 15 percent. This provision mandates a modest "common carrier" function within large systems, without implying that cable systems are common carriers.

• A cable system will not be subject to regulation as a common carrier or utility as a consequence of providing any normal cable service, but it can be so regulated, by state authorities, to the extent that it provides "any communication service other than [normal] cable service, whether offered on a common carrier or private carrier basis." This is the prohibition on unregulated "bypass" which was so avidly sought by the telephone industry.

• Cross-ownership of cable systems by local television stations is prohibited. Ownership by newspapers and other media is allowed under rules to be prescribed by the FCC. Telephone companies (or other common carriers) may not provide cable service except in rural areas and other areas which would not otherwise have cable service. However, a telephone company may construct and own a cable system which it must lease to a separate cable operator or other third party for the delivery of services. A state, or a local franchising authority, is allowed to own a cable system, but not to exercise any editorial control of the content of any cable service, other than programming on an educational or government access channel.

• Pole attachment (payment for use of telephone or power company poles to string CATV cables) is to be state regulated only if a state has issued and made effective rules and regulations for such authority, and takes prompt action on complaints.

• Signal theft is outlawed, with fines of up to $1000 and jail terms of up to six months. A special provision affirms, however, that individuals with private satellite earth stations are allowed to receive unencrypted signals intended for cable operators.

• "Obscene or indecent" programming shall either be prohibited in a franchise agreement, or provided on discretionary channels, subject to the condition that a subscriber be sold or leased, on request, a device, such as a parental control lockbox, to prohibit viewing of that programming during selected periods.

• Subscriber privacy is to be protected through regulation of the collection, use, and disclosure of personal information by cable operators. In particular, "A cable operator shall provide notice in the form of a separate, written statement to each subscriber which clearly and conspicuously informs the subscriber of

(A) the nature of personally identifiable information collected or to be collected with respect to the subscriber and the nature of the use of such information;

(B) the nature, frequency, and purpose of any disclosure which may by made of such information, including an identification of the types of persons to whom the disclosure may be made;

(C) the period during which such information will be maintained by the cable operator;

(D) the times and place at which the subscriber may have access to such information...;

(E) the limitations provided by this section with respect to the collection and disclosure of information by a cable operator and the right of the subscriber...to enforce such limitations.''

In addition, no personally identifiable information concerning any subscriber can be collected or disclosed without the prior written or electronic consent of that subscriber, except to the extent required to render a cable service or other service by the cable operator, or to detect cable theft of service, or to give to governmental entities under court order and under additional restrictions described earlier in this chapter. Disclosure to third party providers of services, upon consent of a subscriber, will not include any description of viewing habits or transactions made through the cable system.

• Affirmative action (equal employment opportunity) programs are to be established by cable operators, and the FCC is to certify annually that each cable system is in compliance with EEO standards. For this purpose, operators must file employment records with the FCC and undergo investigation every few years to evaluate efforts to hire women and minorities. Penalties, including loss of licenses and franchise renewal, are specified.

The outlining of the major provisions of the Cable Act echoes the issues and conflicts surrounding cable services which were the subjects of the earlier parts of this chapter. These provisions are the outcome of uneasy compromises made by the National Cable Television Association, the National League of Cities, the U.S. Conference of Mayors, and the telephone industry following years of bitter contention. The differences between the cable industry and cities have softened in recent years, and the cable industry, founded by daring entrepreneurs, has been in the paradoxical position of seeking government guarantees of franchise permanence and support against encroachments by competing media, especially private cable. Even the provision for leased access, attacking the hallowed concept of operator control of all channels, met with surprisingly muted opposition from the industry. As a columnist in the *Wall Street Journal* put it in the spring of 1984:

> There will be severe pressure on the cable industry to accept the Wirth proposal [H.R. 4103, which developed into the Cable Act] as the price for gaining something close to immunity from competition for its franchises at renewal time. [4]

All in all, the legislation of 1984 responded to the cable industry's requests for more freedom from regulation of pricing and services and more stability in its business environment, while providing at least some protections to franchising authorities, other industries, and the public. The outcome appears to be that civil rights of institutions and the public are protected, while cable operators are freed from pricing restrictions wherever competition in delivery of video services exists. The price paid by the cable industry is that its encroachment onto the territory of common carriers is possibly slowed by the maintenance of the status quo of state regulation. That is, a state may choose to regulate non-cable services such as data transmission. The law rationalized, more or less to everyone's satisfaction, what had been a chaotic and irrational set of institutional and commercial relationships.

The Future of a Medium of Abundance

Cable television has sought to differentiate itself from other media and be freed to develop without constraints through an image as the wide-band highway of the information age and an important provider of public services. In the view of some industry figures, CATV has not received adequate recognition for its public service accomplishments. Trygve

Myhren, Chairman of Time, Inc.'s cable operation subsidiary (ATC), said in 1983 that

> We have to distance ourselves from television...We need to put more emphasis on the fact that we are a significant educational medium with ties to the schools and that we provide services such as C-SPAN that are important to the political process. We are doing all these things that are recognized as significant contributions to the well-being of society, but we're not identified with these things in the public mind. [3]

At the same time, Myhren asserted that "it's been a mistake" to build two-way systems with 100 channels when "50 channels delivered on a one-way system can get the job done." This comment illustrates the industry's ambivalence about its role as the wide-band highway and frustration over governmental pressure for what it perceives as too much abundance. Some late franchise agreements required cable operators to build expensive two-cable, 110-channel interactive systems despite the absence of a relationship between these capabilities and the availability, quality, or salability of programming. Efforts have been made by cable operators in several cities to roll back previous commitments for two-cable residential systems to single-cable systems, perhaps using the new 550 MHz, 57-channel technology mentioned in Chapter 2.

Good programming is scarce and expensive, and although it may not be the industry's fault, this pulling back from the promise of abundance is disappointing, and hopefully temporary. The promise of CATV, other than simple improvement of signal reception, has always been that it would provide so much wide-band capacity that public services could have all they want at minimal cost and even the smallest audience groups would have access to programming to meet their special needs. It used to be believed, on the basis of no experience and wishful thinking, that cable operators could profitably supply programming to meet these needs.

Unfortunately, narrowcasting to audience subsegments has more often than not been an economic failure, unless the subsegments are very large, for example "children," "teenagers and young adults," "sports fans," and "Spanish-speaking." Programming services have tended, as Chapter 3 described, to consolidate into a relatively few mass-audience formats. Furthermore, communities have made very little use of their institutional and local programming facilities, partly because viewers are conditioned by television and movies to expect highly professional video productions on their television screens. Interactive communications has not progressed much beyond the polling applications described in Chapter 4, despite a few small-scale experiments with higher capacity videotex systems. Much of cable television looks like broadcast television because it addresses the same mass-audience tastes with the same type of programming.

But these shortcomings should not dim the original vision of a medium of abundance. Educational programming with broadcast video and two-way audio, especially university courses delivered to off-campus industrial and office locations, has a tremendous potential and can be profitable. Electronic data base publishing will need substantial distribution capacity. Greater efforts at coordination with community, industry, and professional organizations could bring in subsidized narrowcast programming of excellent quality. Finally, technical advances in control of program delivery, especially addressable converters, will be important to developing and exploiting new markets for high-value programming delivered to specialized audiences. The time will come when 110-channel interactive systems, although perhaps not a good present investment for their builders, will not have too much capacity for the growing demands of a truly information-based society.

For Further Reading

[1] I. de Sola Pool, *Talking Back: Citizen Feedback and Cable Technology.* Cambridge, MA: M.I.T. Press, 1973.
[2] George Orwell, *1984.* New York: New American Library, 1983, pp. 169-170.
[3] "Gearing up for the long haul," *CableVision*, June 20, 1983.
[4] "City, cable officials set for new cable bill talks," *Multichannel News*, Apr. 16, 1984.
[5] "Cable television," FCC Info. Bull. #18, Mar. 1982.
[6] G. M. Hauser, Address to Women in Media Conference, Annenberg School, Philadelphia, PA, Sept. 30, 1982.
[7] Copyright Act of 1976, Public Law 94-553, Title 17 U.S.C.
[8] "Foundation challenges CATV franchise exclusivity," *Private Cable*, Mar.–Apr. 1983.
[9] "The emergence of pay cable television," report prepared for the NTIA by Technology + Economics, Inc., Cambridge, MA, vol. I, p.7, July 1980.
[10] W. Buffett, "When cable TV comes to town," *Washington Post*, Sept. 7, 1980.
[11] "Cable TV bidding war to serve large cities is quickly cooling off," *Wall Street Journal*, Apr. 1, 1983.
[12] "The 'smoking gun' case," *CableVision*, June 13, 1983.
[13] "Interactive home media and privacy," Collingwood Associates, Inc., Washington, DC, report prepared for the FTC, Jan. 1981.
[14] "Two decisions cause confusion on cable regulation," *The New York Times*, July 19, 1984.
[15] *Cablevision*, Aug. 6, 1984.
[16] "Will high court ruling vex must-carry crusade?" *CableVision*, July 9, 1984.
[17] "Taking the first," *CableVision*, May 7, 1984.
[18] "Cable TV notes," *The New York Times*, Feb. 19, 1984.
[19] Jack Anderson's syndicated column appearing in newspapers June 26, 1984, e.g., *Washington Post*, p. C13.
[20] "Bitter fight endangering cable TV bill," *Washington Post*, May 19, 1983.
[21] *CableVision*, Apr. 16, 1984.
[22] "Accord brings hope for cable television bill," *The New York Times*, Sept. 29, 1984.
[23] "Pole rate debated," *CableVision*, Sept. 19, 1983.
[24] Cable Communications Policy Act of 1984, U.S. Government Printing Office.
[25] "Free shares of cable TV cost its users," *Washington Post*, Sept. 14, 1980.
[26] T. Whiteside, "Onward and upward with the arts: Cable 1," *The New Yorker*, May 20, 1985; for subsequent articles in the series see May 27, June 3 issues.

6 The Shape of the Competition

Urban cable systems may hold exclusive operational franchises, but they have no monopoly on communications. In a communications industry remarkable for its diversity, growth rate, and importance in national affairs, it would be surprising if urban cable systems had the unique capability to provide any communications service. The fact is that the cable industry faces competition in every part of its business. For any given cable service, there is at least one alternative medium, perhaps insignificant now but likely to become important in the future, which is arguably just as good a delivery system. As Senator Bob Packwood said in early 1983, with direct broadcast satellites in mind, "If you are not already in cable, I would not advise you to go after a major franchise. It wouldn't make sense to string all that wire when these little dishes are coming" [1].

Cable will have to share its market with these alternatives, and in some cases accept less than the lion's share of the available business, but the competition in video programming is not likely to be devastating. CATV is in a strong position, with a large installed base and tens of millions of customers. Moreover, competition could be a healthy spur to innovation and efficient operation, and, by demonstrating that cable is not a monopoly industry, will help hold off the reimposition of regulatory constraints.

The competition does, unfortunately for cable, pose a threat to its profitability. If competitive video delivery media prevent franchised operators from reaching and holding the 50 or 60 percent penetration necessary to recoup their large capital investments, investment in cable will become unattractive and cable systems will have to scramble to remain profitable at all. Cable operators may do best in the long run by turning to the development and packaging of information, transaction, and entertainment services, with little preference for coaxial cable over other distribution media. One possible scenario, which appears several places in this book, is the transfer of cable services to a switched optical fiber network which might be built in the 1990's by the telephone industry (see Chapter 12) or some still undefined financing entity.

The alternative transmission technologies—cable's competition—may be assigned to four broad categories (Table 6.1):

1) Distribution of video programming (especially pay TV).
2) Low- and medium-rate interactive services such as security systems, environmental management, limited information retrieval, and home banking.

Table 6.1: Four categories of communication services and the media which can deliver them.

Service Category	Usable Delivery Media
Distribution of Video Programming	Coaxial cable system "Free" television broadcasting, including low-powered "drop-ins" Subscription television (STV) Multipoint distribution service (MDS) Direct broadcast satellite (DBS), including satellite master antenna television (SMATV) Videocassettes Subscriber-switched optical fiber
Narrow-band Interactive Services	Coaxial cable system (with refinements) Telephone network, present and with data network enhancements Subscriber-switched optical fiber
Wide-band Interactive Services	Coaxial cable system (with refinements) New regional and metropolitan wide-band networks Microwave distribution such as digital termination service (DTS) Subscriber-switched optical fiber
Local Telephone Services	Coaxial cable (with refinements) Telephone network New regional and metropolitan wide-band networks Subscriber-switched optical fiber

3) Higher rate interactive services, oriented toward business users, such as high-speed data transmission and videoconferencing.
4) General telephone services, meaning normal voice and data services similar to those currently provided by telephone companies.

Although an urban cable system, suitably refined, can theoretically provide all of these capabilities, the new investment and building required to do so has only rarely been justified. It may not, for example, be economical to upgrade older one-way entertainment distribution cable systems to provide interactive services, and it is doubtful that cable will ever take more than a tiny share of the residential telephone business.

Bulk business-oriented transmission services are more lucrative and relatively easy to do, but the cable industry has reservations about these also. John Gault, the president of Manhattan Cable, a company active as a carrier of high-speed business data traffic, stated in 1983 that commercial data transmission on cable is "still a high-risk business" [2]. Operators of competing distribution services have, not surprisingly, an even less optimistic view of CATV's relative competitive merits. Steve Wechsler, chairman of the board of Marquee Television Network, an MDS operator, was quoted as follows:

> There are a lot of other technologies which can do the same thing that cable can do. For instance, the phone company with deregulation is going to provide security, banking, shopping and data transmission a lot better than the cable companies. You can rent a videocassette now for 99 cents, and that's probably more convenient and less annoying than repeats on the pay TV channels. [4]

"A lot better" can have many meanings, but the principle physical and psychological shortcomings of the cable industry are linked to its origins as an entertainment rather than a communications industry. Short on engineering staff and having neglected, in many cases, to spend as much as it should have on quality and reliability in its plant, the CATV establishment is not always in a position to take the next technical step.

In the face of its own doubts and discouragement from outside, CATV will have a tough fight to expand its communications services and hold its market share in video entertainment. But cable has three substantial advantages over most of the competitors: it is in place in most populated areas of the United States; the investment to put it in place has been made on the basis of residential subscription revenues alone, so that any new uses of the cable require only an incremental investment for upgrading; and it can deliver many channels with relatively inexpensive subscriber electronics. So perhaps the fight is a fair one. The remainder of this book, an introduction to the (sometimes) competing media, makes only a cursory effort to predict the outcome of the competitive struggle, but tries to bring out the technical and business challenges which the competitors pose to CATV.

The relationships between CATV and other media will not always be antagonistic. Beyond their competitive potential, the alternative communications media can join with cable to offer specific services which are difficult to provide on one medium alone, and they are elements, together with cable, in the more integrated communications environment toward which we seem to be heading. No major contemporary communication medium is likely to disappear. Instead, competitive pressures will help it find its economic niche doing what is most logical and cost-effective for it to do.

The major alternative media are described in some detail in the chapters following this one. This introductory chapter provides an overview and the basic definitions, looking at the competition one service area at a time: video programming, narrow-band interactive services, wide-band interactive services, and local telephone service.

Video Programming

Important as the technical differences among wide-band delivery media may be for consumer acceptance, they are overshadowed by the competition for the most popular entertainment and the clashes of interests along the distribution chain, some of which were described in Chapter 3. The producers of programming will probably have the last word about distribution policies, based in large part, of course, on the changing whims of the consuming public.

In 1982, as Table 6.2 shows, cable got the bulk of consumer expenditures for home video entertainment. By 1984, despite a large growth in the cable subscriber population, the cable industry had not significantly increased its market share. Sales and rentals of prerecorded videocassette tapes, although still relatively small compared with cable revenues, had suddenly and unexpectedly become the principal threat to pay cable. Videodisks, a forecaster's favorite of earlier years, had for all practical purposes dropped out of the market.

Direct broadcast satellite service (DBS), which did not exist commercially in 1982 and had barely appeared, in a compromise form, in 1984, could make significant inroads on cable in the latter part of the decade. There are several kinds of direct satellite broadcasting, and it is easy to become confused about what this competition really is. The more than one million

Table 6.2: 1982 and 1984 distributions of expenditures for home video equipment and programming (not including computers and videogames).*

| | Millions $ | |
	1982	1984
STV	415	138
MDS	177	88
Blank videocassettes	280	720
Prerecorded videocassettes (sales)	350	1000
Prerecorded videocassettes (rentals)	400	900
Videodisks	100	120
Pay cable	2400	5000
Basic cable	2800	4300
TV sets	5000	6130
VCR's	1280	3100
Videodisk players	54	55
Video cameras	230	460

*The totals for 1982 and 1984 were $13.5 billion and $22 billion, respectively. Motion picture box office receipts, for comparison, were $3.5 billion and $3.9 billion in 1982 and 1984, respectively. (Sources: "Home entertainment dollar," LINK Information Resources, 1984; *Multichannel News*, May 21, 1984; The Pay TV Newsletter, Paul Kagan Associates, Jan. 22, 1985; *The New York Times*, Mar. 3, 1985; *Billboard*, June 30, 1984; "Home video & cable report," Knowledge Industry Publications, 1982).

American owners of backyard satellite antennas intercepting programming intended for cable operators are enjoying an unofficial, but legal, direct satellite service, although they are already being faced with scrambled signals and demands for payment. These private receiving stations still cost in the thousands of dollars, but are much less expensive and more capable than they once were.

What DBS is really intended to be is a new satellite service using high-powered satellites which broadcast signals strong enough to be received by small and inexpensive private earth stations (Fig. 6.1). A cost of around $300, and an antenna diameter of about 3/4 meter, are often suggested for equipment which could be mass-marketed to ordinary television viewers. High-powered DBS also intends to deliver signals of significantly higher quality than those available to most cable subscribers, even without introducing new enhanced definition television formats different from the NTSC broadcasting standard. This emphasis on signal quality could pose a threat to the cable operators who do not make the effort that they should to deliver signals of superior quality.

The prospective American operators are each planning systems of two to three satellites to cover separate service zones of the United States, and European DBS systems are being designed which will give each country control over a few high-powered satellite-borne transmitters beaming programming to its national territory. Each U.S. DBS system is expected to offer its service regions an initial minimum of six channels, with the possibility of many more in the future from colocation of satellites and use of spot beams, as described in Chapter 10.

Fig. 6.1: Direct broadcast satellite (DBS) service from high-powered satellite-borne transmitters to inexpensive residential receiving station.

A compromise and perhaps interim DBS system, using a more economical medium-power satellite for broadcasts to moderately priced earth stations, has also been of interest, and a commercial service did in fact begin in 1983. A medium-power system operates with a smaller safety margin against noise

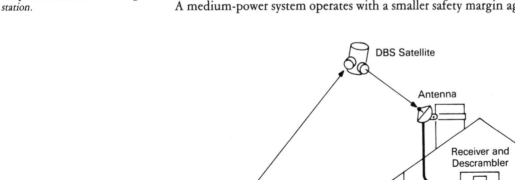

and a more expensive earth station than the full-powered DBS service, but may be good enough to discourage early development of the extremely expensive high-powered satellites.

Private cable, which is a cable system limited to an apartment complex or hotel and is frequently in competition with the franchised cable operator for subscribers, lies somewhere between DBS and CATV. It is often called satellite master antenna television (SMATV, Fig. 6.2) because the old master antenna system has been replaced (or augmented) by a full-sized TVRO earth station, receiving some of the same relatively low-powered satellite broadcasts which supply commercial cable operators. The substantial cost of the TVRO earth station, perhaps $15,000, is shared among the subscribers. As long as the mini cable system does not cross public property, its operators can maintain that it is not, unlike a franchised cable operator, subject to regulation. SMATV is an immediate threat to the profitability of urban cable systems, since it skims off the cream of easy-to-wire apartment complexes.

Satellites are being planned which are intended, in part, to service the medium-powered DBS and SMATV markets. RCA, for example, in a plan

Fig. 6.2: *Satellite master antenna television (SMATV) service, a private cable system with a relatively expensive satellite receiving station.*

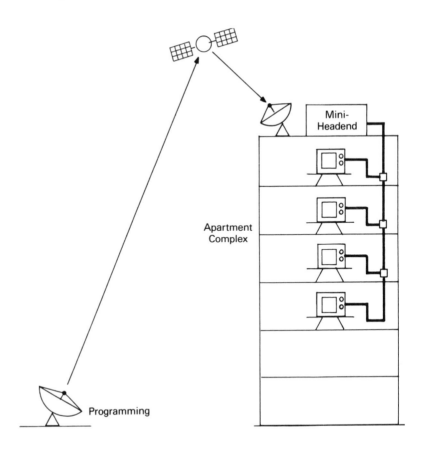

filed with the FCC in 1983 for three satellites to be launched between 1989 and 1992, emphasized the impetus the satellites would give to the new broadcasting services, including private cable.

Subscription television (STV, Fig. 6.3) is one of cable's older competitors. It is simply the dedication of part of the operating time of a television broadcast station, usually UHF, to subscriber-paid programming. During the hours of broadcasting, the television signal is scrambled and appears as a distorted jumble on television screens of non-subscribers. The STV operator provides a descrambler to paid subscribers which clears up the picture.

An STV station broadcasts only one channel of programming, and its access to the public is limited to the station's normal over-the-air reach. This is a serious disadvantage in competition with CATV, which usually drives STV out when it arrives in a previously uncabled area. STV does, however, have the substantial advantage of immediate setup, requiring only minimal alterations at the broadcast station and placement of moderately-priced ($100–150) descramblers in subscribers' homes. The FCC's authorization of many new low-powered television (LPTV) stations opens the possibility of multiple STV channels in some service areas, although the low-powered stations are intended primarily for scattered deployment in underdeveloped service areas. It remains to be seen if LPTV can revive the shrinking STV market.

Multipoint distribution service (MDS, Fig. 6.4) is, like subscription television, an older service but with more expansion potential. MDS is a terrestrial microwave broadcasting system in which the operator sets up an omnidirectional transmitting antenna on some high point in direct line of sight from a large number of residences, and leases inexpensive (about $150) microwave receiving stations to subscribers. Like STV, MDS has the advantage of quick and inexpensive setup. Although most MDS services now provide only one or two channels, the FCC's recent opening up of channels in another microwave band, that of the instructional television fixed service (ITFS), could increase the number of channels to as many as 12 in a single metropolitan area. This new multichannel multipoint distribution service (MMDS) could provide a program package meeting most

Fig. 6.3: Subscription television (STV) on an over-the-air broadcast signal.

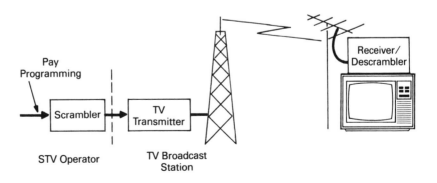

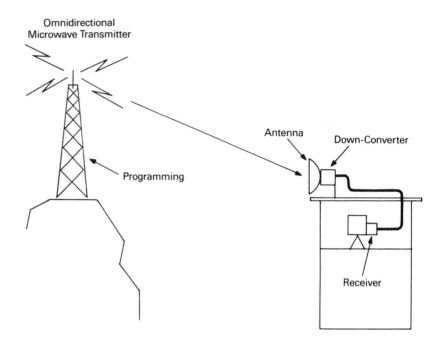

Omnidirectional
Microwave Transmitter

Antenna

Down-Converter

Programming

Receiver

Fig. 6.4: Multipoint distribution service (MDS), a line-of-sight terrestrial broadcasting system used for both pay services and business communications.

subscriber needs at a cost of \$175 to \$250 for each subscriber's receiving equipment, and pose a threat to cable systems with a much greater per-subscriber capital investment to recover. MDS does, however, suffer from its line-of-sight requirement and an effective range of less than 20 miles.

Over the air free *television broadcasting* is, of course, a competitor of cable, even though it is carried by cable. Broadcast programs compete with CATV pay and advertiser-supported channels for audience share, and it is those channels which bring in the big revenues for cable operators. Cable appears to be slowly gaining in this competition, as was noted in Chapter 3, but the audience share of broadcast television may stay relatively constant for some time.

Broadcasters have even considered encroaching on CATV's pay territory beyond their existing minimal STV activity. In January 1984, ABC launched a service called "TeleFirst" in Chicago which broadcast scrambled signals of current movies, through an ABC affiliate station during noncommercial broadcast hours to specially-equipped home VCR's. The installation fee of \$75 and monthly rates between \$18 and \$26 were apparently too high for TV viewers in Chicago, because TeleFirst attracted only a few subscribers and was abandoned in June of that same year. Arthur Cohen, president of TeleFirst, blamed its demise on the competition from prerecorded videocassette tapes.

This book has a "new media" scope and does not examine broadcast television as a CATV competitor beyond the few comments here. Broadcasters are, of course, heavily involved themselves in the new media, especially as cable operators. The possibility also exists for the national broadcasting networks to become program distributors by satellite to cable systems, as they already are to their own affiliates. This would further blur the already vague lines separating the broadcasting, satellite distribution, and cable industries.

Videocassette recorders (VCR's) have proliferated at an amazing rate in recent years, and stores renting prerecorded videocassettes (Fig. 6.5) have become the major competitor to CATV for pay programming revenues. It is somewhat ironic that in the age of telecommunications, a distribution medium which requires a personal appearance by the customer at a retail store should be so successful. This is a clear message from consumers that they want the choice and viewing time convenience which they have not gotten from CATV and are not likely to get until (and if) video on demand becomes a reality.

VCR's were initially complementary to, rather than competitive with, CATV, being used almost exclusively for "time translation" of broadcast programming and for viewing X-rated materials. Off the air (or cable) recording is still the heaviest use of VCR's and the target of legal and legislative attacks from movie studios claiming copyright infringement. But the relationship seems to be turning upside down as prerecorded videocassettes become a threat to the cable industry and an opportunity for studios as a lucrative first-run consumer outlet. The cable industry is justifiably disturbed by the videocassette competition, although it has distinguishing capabilities for live broadcast of sports and entertainment, including pay-per-view presentations, which industry leaders feel may limit the VCR inroads.

Subscriber-switched optical fiber (Fig. 6.6), in which each subscriber has an individual fiber link direct to a wide-band digital switch, is a possibility for near-universal home communications about the end of the century. It

Fig. 6.5: A retail outlet for rental and purchase of prerecorded videocassettes, the fastest-growing video distribution medium of the mid 1980's.

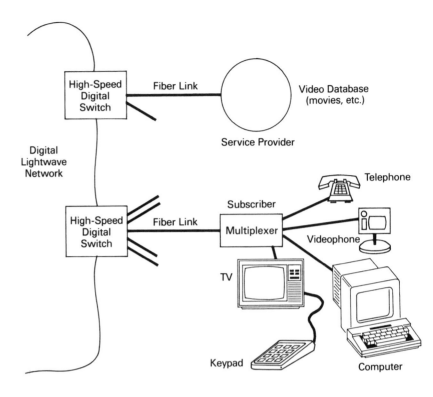

Labels in figure:
High-Speed Digital Switch — Fiber Link — Video Database (movies, etc.) — Service Provider — Digital Lightwave Network — High-Speed Digital Switch — Fiber Link — Subscriber Multiplexer — Telephone — Videophone — TV — Keypad — Computer

Fig. 6.6: An optical fiber-based wide-band switched communications network would allow almost unlimited subscriber choice in video entertainment and sophisticated communications services "on demand."

would offer the subscriber a range of choice in wide-band services impossible on any broadcast system, including personal movie selection and two-way videotelephony. It is generally acknowledged that a network of this kind would obsolete cable systems as they are conceived today.

The telephone industry is already developing the technical components of this dramatically advanced future network, and examining the services which it might deliver. Video entertainment on demand (within certain limits) is one of these services, and a case could be made that the fiber network is justified only if video entertainment services migrate to it from cable.

Prior to this long-range outcome, which is likely, but not certain, cable systems could become integrated with optical fiber trunking facilities in wide-band regional networks with limited subscriber switching capabilities. Individual subscribers would not have private lines into video switching centers, but CATV hubs higher up the distribution tree could be integrated with telephone company wide-band switching centers, providing more demand-assignable communications capacity for subscriber services, including limited video on demand.

There is some possibility that another utility might build municipal optical fiber networks, perhaps compromising between tree-and-branch and fully switched architectures. Power companies are possible candidates.

Existing cable companies would surely challenge this competition as a violation of their franchise agreements.

These competitive delivery systems, with the notable exception of switched optical fiber, transmit one way only in residential applications, which is good enough for video entertainment. Interactive data communication could be arranged in hybrid systems with the telephone network, just as it can for interactive cable. Although even collectively these alternative wide-band delivery systems do not now have anywhere near the subscription base of cable, they are capable of making substantial inroads.

The relatively small capacities (number of channels) of many of the competitive video media are often pictured as a fatal disadvantage with respect to cable, but not everyone agrees that a large number of channels is needed to attract viewers. A few very good channels, or a wide selection of materials available with some delay as with prerecorded videocassettes, might be sufficient. An era of competition is being shaped, with major shakeouts possible in the future, but with the near-term viability of all contending video media encouraged by an open and still immature consumer market.

Narrow-Band Interactive Services

Cable systems built to distribute video entertainment are not very well designed for the narrow-band services described in Chapter 4, such as security, energy management, and low-to-moderate-rate information services. On the other hand, neither is their only serious rival, the telephone network, which still, at most subscriber locations, accepts data traffic only if it is first transformed by modems into signals which can pass through a voice channel.

The telephone network offers a variety of interactive data transmission facilities, ranging from ordinary dialed telephone lines to high-speed leased digital circuits. For the residential user, the only economically feasible choice in most locations is a dialed telephone line [Fig. 6.7(a)], with the data rate limited to 1200 bits per second by modem cost. This is adequate for ordinary information services, but not good enough for videotex browsing or rapid delivery of highly detailed displays. Cable systems, in contrast, can operate at the higher data rates needed to deliver highly detailed visual "frames" in a very short time. A further limitation of dialed lines is the setup, or "getting through" time.

Why, then, is the telephone network such a formidable competitor for narrow-band data services, and already so widely used for data communications? There are three answers: first, the telephone network is universally interconnected; second, the telephone network is far ahead in actually offering data services; and third, the telephone network is planning major improvements to its data communication facilities and has the engineering resources to make them happen.

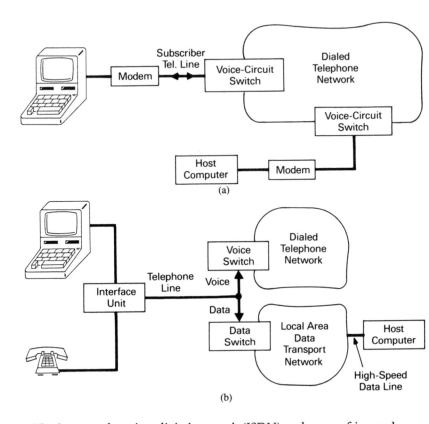

Fig. 6.7: Narrow-band interactive data communications in the telephone network. (a) Present capability for data transmission through the dialed voice network. (b) LADT (local area data transport), in which a digital transport network is used instead of the voice telephone network.

The integrated services digital network (ISDN) and some of its predecessor facilities, particularly LADT [Fig. 6.7(b)], will gradually bring the digital access point out to subscribers and meet most service provider and user needs for fast call setup, adaptable (and higher) data rates, and cheap data transport. Standard user interfaces will allow the connection of data terminals and other subscriber equipment without the need for modems, allowing higher performance in low-cost terminals. Just how quickly this will happen depends on technical and regulatory events which are hard to forecast, but there is going to be a great deal happening in the next several years. CATV has only a brief time in which to establish a competitive foothold. There are indications that many cable systems will forego data communication and join in hybrid systems with the telephone network. Cable will only be able to compete in narrow-band interactive services beyond program selection if it develops data services more quickly and competently than it has in the past.

Wide-Band Interactive Services

The boundary line between ''narrow-band'' and ''wide-band'' interactive services is arbitrary, but in data rates, 64,000 bits/second might be the lower edge of the "wide-band" category. Moving video, whether sent in analog or digital form, is clearly wide-band traffic.

Wide-band interactive services should properly be understood as wide-band in and out of a subscriber location, but it is often taken to mean wide-band transmission downstream to a subscriber coupled with lower-rate interactive data transmission. With this looser definition, it is conceptually no more difficult for the competitive wide-band media to be made interactive than for cable itself. STV, MDS, DBS, and SMATV can all hook up with the telephone network in hybrid systems, e.g., the satellite service shown in Fig. 6.8, to become interactive. Interactive DBS, with video broadcasting to remote viewing locations and two-way voice carried through the telephone network, is offered as a videoconferencing service by major hotel chains. There is no reason why residential DBS or other wide-band media could not be interactive in the same way, given a good reason for making them interactive. There are even possibilities for building interactive capabilities directly into these media. As one early indicator, the FCC in 1984 granted Equatorial Communications Company authorization to deploy transmit and receive earth stations, with four-foot antennas, for two-way data communication at rates up to 9.6 kbits/second.

Interactive services which are wide-band in *both* directions are intended mainly for major business users. Cable systems are regarded by some as a

Fig. 6.8: A hybrid interactive system for direct broadcast satellite (DBS) service. Some videoconferencing services offered by hotel chains are exactly this system.

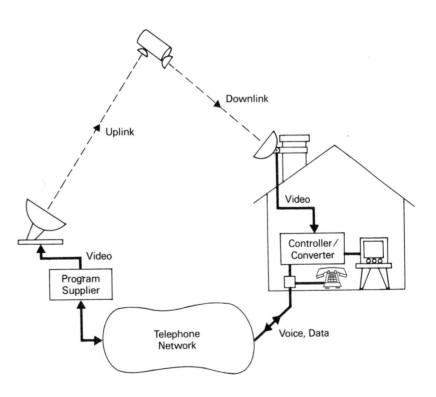

major sleeper among the potential bypass services, with the possibility of taking considerable business away from telephone companies, which have recognized the threat and fought back (see Chapter 5). Cable's advantage is a wide-band medium with a seemingly low incremental cost of adding point to point high-speed transmission services. CATV systems do, however, have some serious competitive disadvantages with respect to telephone companies, including fragmentation of urban areas among different operators, inadequate wiring of business districts, greater susceptibility to service interruptions and, as noted earlier, weaker engineering capabilities.

The telephone network will become an even more formidable competitor with the development of the integrated services digital network, and a devastating one if and when it builds a wide-band distribution network. If subscriber-switched optical fiber should become a reality, cable would become obsolete for almost all services.

In addition to this potential competition, cable is also faced with competition from radio services. Digital termination service (DTS), a high-speed two-way data service with a central transmitting-receiving station and small subscriber transmit-receive microwave stations, may also contend for business data traffic. DTS can offer data rates up to several megabits per second at costs which could potentially be very low. Multipoint distribution service (MDS) is a long-established microwave service which could also develop business-oriented capabilities.

Local Telephone Services

Chapter 4 described early efforts by Cox Cable Communications to carry bulk voice communications from business customers to interexchange carriers and Chapter 5 examined the regulatory controversies. This was a direct challenge to regional telephone companies in access services for interchange carriers, one of their most profitable businesses. It does not look as if cable operators want to make this a serious challenge.

These bypass interests extended also to residential telephone service, and experiments were done on dialing through the cable to the switches of interexchange carriers. There were significant difficulties, including the privacy of conversations and the expense of the required subscriber modulation equipment. Further significant development is unlikely here also.

The Competition Integrated with Cable

As a video broadcast medium, cable has a large capacity, a relatively straightforward technology, and a market penetration far exceeding any other medium except broadcasting. The competing delivery media may take away enough of the video entertainment business to reduce profits to modest levels, but cable has the potential to get some of the profits back by providing ancillary communications services.

The competing media will also work together with cable. Direct satellite broadcasting will supply network programming and other services to cable systems and will be sold by cable operators to their own fringe areas. VCR's will continue to be used much of the time as "time translators" for cable programming and may eventually be used for video-on-demand delivered through cable systems. Cable systems will interconnect with each other through satellite, microwave, optical fiber, and telephone line communications, some of them provided by competing industries. Most of all, hybrid systems with the telephone network, making the most of cable's wide-band downstream capabilities and the telephone network's narrow-band interactive capabilities, will help cable realize the interactive services which have proven to be so difficult to create entirely within the cable environment.

The telephone network will probably dominate the data transmission market, but cable systems may win pieces of the business for transactional communications, high-speed data transmission within cities, bulk business traffic, and possibly some residential data traffic. The few examples to date of local communication services in cable systems are milestones in the development of cable and its integration into the national communications network, but not indicators of a major industry advance.

For Further Reading

[1] "Competition in cable," *The New York Times*, Mar. 7, 1983.
[2] *Fortune*, Apr. 18, 1983.
[3] *Channels of Communication 1983 Field Guide.*
[4] *Cablevision*, Aug. 6, 1984.
[5] T. E. Bell, "The new television: looking behind the tube," *IEEE Spectrum*, vol. 21, pp. 48–56, Aug. 1984.

7 Subscription Television

It has been apparent almost from the start of television broadcasting that a television station could transmit pay as well as free video programming, and subscription television (STV) is simply a one-channel video pay service offered through a television broadcast station (Fig. 7.1). The stations carrying STV are invariably in the ultra-high frequency (UHF) band, and usually transmit scrambled pay programming in the evening. No major capital expenditures are necessary other than for the subscribers' descramblers. Although subscription television has, in the words of an FCC official, "pretty much dropped dead" [11] as a consequence of the spread of CATV, it has some potential for rebirth in low-powered television (LPTV). It can also be viewed as an early model for future broadcasting services with time-shared channels.

An STV operation is handled by an STV franchisee, who is often (but does not have to be) the station licensee. A franchisee is responsible for all aspects of the subscription service, including securing programming, maintaining subscriber equipment, and billing for service. The franchisee may be a program distributor as well.

STV is not as old as cable. In 1957, the FCC, which regulates television broadcasting, decided that it had statutory power to authorize the operation of subscription television services, and proposed that trials of pay television should be made. It proclaimed rules for a limited trial of these services, restricting experiments to cities with at least four conventional TV stations and requiring a minimum number of hours of free, non-scrambled programming during the broadcasting day. Only one STV station was to be allowed in each city.

There was opposition from theater owners, who found free television to be enough competition, and this, together with bureaucratic requirements and startup problems delayed the beginning of STV operation until the summer of 1962, when UHF (ultra-high frequency) station WHCT in Hartford, Connecticut, began transmitting pay programming. Based on three years of operation of WHCT which the FCC regarded as successful and not damaging to "free" broadcast television, as well as on other information, the FCC decided in 1968 to go ahead with nationwide authorizations for new STV stations.

At first, restrictions similar to those for the experimental period were imposed: STV was to be allowed on only one station and only in

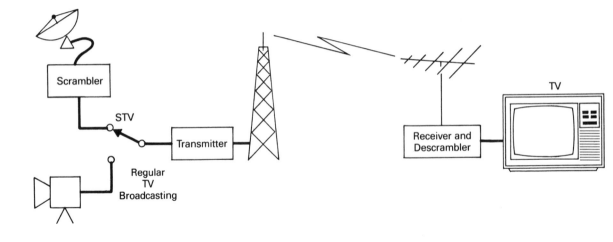

Fig. 7.1: Subscription television (STV) on an over-the-air UHF television signal.

communities having at least five commercial television stations; the STV station would still have to broadcast a minimum of 28 hours per week of "free" programming; some types of programming could not be siphoned away from "free" television; and the movies shown could not be more than two years old.

A court challenge by the National Association of Theater Owners and the Joint Committee Against Toll TV was unsuccessful, and STV was off and running. The first formally authorized STV station, the present WWHT in Newark, New Jersey, began operation in 1977. The "antisiphoning" rule was dropped in 1977, after a legal suit won by HBO with regard to cable programming, and the "one-to-a-community" rule was rescinded in September 1979, although the limitation to communities with at least five stations (including the STV station) still stands. There are no longer any limits on broadcast hours, and the FCC, regarding STV as a programming package rather than a broadcasting operation, has lost interest.

Between 1969 and 1982, roughly 40 STV stations were authorized by the FCC, and 33 were in operation by the end of 1982. This number was reduced to 12 by 1985, listed in Table 7.1.

Despite its low per-subscriber equipment costs of about $200, a negligible setup time, and a lot of early enthusiasm, STV has not done well. Its main problem is that an STV station transmits only one channel, in contrast to the multichannel capacity of a cable system. In addition, STV operators, perhaps because of operating inefficiencies connected with their relatively sparse and scattered audiences, have demanded subscription rates comparable to cable, in the $20–25 range, with an additional $5–10 in many STV operations for a second tier of mostly adult programming. "In cases where cable comes in, subscription TV can lose one out of every two customers," according to industry analyst Alan Cole-Ford in 1983 [2]. Raymond Peirce, president of Oak Industries, a major STV operator, conceded at that time

Table 7.1: STV operations on the air at the beginning of 1985.*

Market	Station	Channel	STV Service (Franchise Holder)	Subscriptions
Los Angeles	KBSE	52	ON TV (Oak), replaced by SelecTV in mid-1985 STV subscribers moved to KWHY in late 1985.	147,000
Los Angeles	KWHY	22	SelecTV of Los Angeles	49,000
Chicago	WSNS	44	ON TV, replaced by SelecTV. STV ended in 1985.	64,500
New York	WWHT	60	Cooper Wireless Cable	59,500
Washington/ Baltimore			Subscription Television of America	41,000
Boston			Subscription Television of America	18,500
Philadelphia			Pay TV PA/PRISM	14,500
Ann Arbor			IT Subscription TV	13,000
Cincinnati/ Dayton			ON TV (until 1985) SelecTV	6000
Reading, PA	KTMA	23	STV Reading	4500
Minneapolis			Spectrum	1000
Norwood, MA	WTVA	51	Preview	—

*Sources: The Pay TV Newsletter, Paul Kagan Associates, Inc., Jan. 22, 1985; SelecTV of California, Inc.

that "we haven't been able to come up with unique programming of sufficient frequency and sufficient interest to boost subscribership."

By the third quarter of 1983, the total number of subscribers to STV services had dropped to 900,000 from a 1982 peak of 1.4 million, and large operators were leaving the industry. The slide has continued, down to 430,000 subscribers at the end of 1984 (Fig. 7.2), but it may be too early to assume that STV is doomed to extinction. Efforts could be made to increase the number of STV signals available to a subscriber by coordinating the activities of several operators in the same service area, assuming, of course, that there *are* several operators. The FCC's authorization of a large number of new low-powered television stations (LPTV), described later in this chapter, may provide new broadcasting outlets for STV operators. The FCC has also tried to aid the STV industry by exempting STV-carrying stations from some of the transmission, licensing, and station identification standards imposed on ordinary broadcast television stations.

STV operators may hold on to some carefully selected and cultivated markets. By concentrating on areas with the proper geography for good UHF over-the-air reception, on areas without CATV or the near-term prospect for CATV, and on subscribers affluent enough to spend $20–30 each month for a single pay channel and unique programming for those subscribers, STV may stay in the race. There are possibilities for symbiotic relations with cable

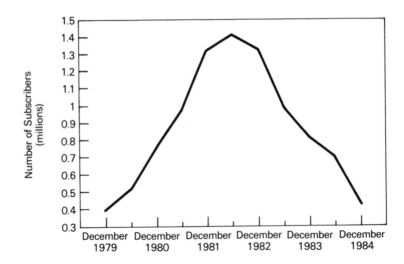

Number of Subscribers (millions)

December 1979 | December 1980 | December 1981 | December 1982 | December 1983 | December 1984

Fig. 7.2: The growth and decline of STV subscriptions. (Source: The Pay TV Newsletter, Paul Kagan Associates, Inc.)

systems in which STV, along with other over-the-air pay systems, would fill in those areas which are not economical to wire. In this situation, an STV station could be supplied programming from the cable headend.

The largest past STV operation was National Subscription Television (NSTV), a partnership of Oak Industries and the Chartwell Communications Group. In 1985, NSTV sold its major outlet, KBSE-TV, which is channel 52 in Corona, California. This channel began STV operation in 1977, and in 1982 had about 380,000 subscribers. NSTV's pay programming package, called "ON TV," which offered movies, entertainment specials, a monthly cultural program, and local sporting events, went out of business in 1985, in part reflecting the greater availability to non-CATV outlets of the major CATV programming services. Wometco Home Theater, SelecTV, Subscription Television of America (STVA), Satellite Syndicated Systems (SSS), and Home Entertainment Network (HTN) were the remaining STV programming services active at the beginning of 1985.

How it Works

The UHF broadcaster, with an operating range of about 20 miles around the transmitter, can carry pay programming supplied by satellite, microwave, or tape. The pay signal is scrambled by one of the conventional techniques described in Chapter 2, and descrambled in a box (Fig. 7.3) provided to the subscriber. Pay-per-view might have been a profitable alternative to monthly subscriptions, but only a few STV operators installed the addressable subscriber equipment, and they only occasionally secured the first-run movies and sports events which were needed. A number of the STV operators had, as suggested earlier, two-tiered or even three-tiered

services in which a "basic" subscriber received only part of the daily schedule of pay programming.

Aside from their disastrous loss of subscribers and legitimate operational difficulties, STV operators have had a serious signal theft problem. Retail vendors of descramblers argued that there is nothing illegal in detecting and decoding signals sent unsolicited through the air, and the courts supported this view. In August of 1980, a United States District Court Judge ruled against the nation's largest STV company, National Subscription Television, in its suit against several electronics companies and individuals who had been selling descrambling kits. The kits sold for about $140; some electronics stores gave customers a list of people who would install the decoders.

Cable carriage of STV signals has been another major controversy of this industry. Cable operators are, of course, extremely reluctant to carry a competing pay service for free, but STV feels it is unfairly cut out of the market when cable subscribers disconnect their receiving antennas. STV operators tried to convince the FCC that their scrambled signals should be carried by cable systems just as the local "free" broadcast signals which cable operators were once compelled to carry. Wometco Blonder-Tongue Broadcasting Corporation, the licensee of WWHT-TV in Newark, New Jersey, petitioned the FCC in 1980 to require mandatory cable carriage of local STV signals, and was turned down with the argument that arrangements for cable carriage can be made in the marketplace. National Subscription Television, for example, made this arrangement in Los Angeles by sharing profits with the cable operators.

Low Powered Television (LPTV)

One hope for a revival of STV service is in the thousands of new low-powered low-cost VHF or UHF television stations ("drop ins") which the FCC envisioned in its Notice of Proposed Rulemaking of September, 1980. This Notice

proposed to authorize a new broadcast service consisting of very low powered television stations. The principal rule change proposed was that TV translators, permitted only to rebroadcast simultaneously the programming of a full service station, would be allowed to originate programs to an unlimited extent and/or conduct subscription TV (STV) operations. [5]

These low-powered stations could be built for about $100,000, rather than the one to two million dollars for a conventional TV station, and would have a broadcasting reach of only 10 to 20 miles.

The new service was expected to be a boon to smaller towns, ethnic communities, and cultural interests underserved by normal television stations, and when the FCC started accepting applications for licenses in early 1981, its chairman, Charles Ferris, called LPTV "the first new broadcast service in twenty years offering the same intriguing possibilities as the advent of commercial television broadcasting in the late 1940s" [6]. The response was immediate and huge. "It was Charlie's last and greatest populist scheme," one of his fellow FCC members said. "I don't think any of us realized so many business interests would try to get in on the gold rush—or that it would turn out to be such a bureaucratic disaster" [7].

Interest in the new service came from every variety of existing business and entrepreneur, ranging from the smallest independent individual to corporate giants including Sears, Gannett, Federal Express, ABC, and NBC. The FCC was swamped with applications, more than 12,000 of them, and in April, 1981, declared a freeze on applications and a postponement of licensing until it was able to find a way to deal with this almost impossible burden. An exception to the freeze was made for applications proposing to serve areas with fewer than two full service TV stations or to change the frequencies of existing translators. Many of the smaller applicants, and the community groups they were planning to serve, were severely disappointed by the licensing delay, which for some localities may be years. The first application was granted in May, 1981, and 300 were granted by September, 1983, most of them in Alaska.

The FCC first approached the licensing problem by deciding to accommodate the most rural areas first. The rural classification is in three "tiers," based on ranking TV markets by potential audience size. Tier I applications were defined as those proposing to locate the transmitting antenna more than 55 miles from the 212 highest ranking cities, Tier II as those with transmitter more than 55 miles from the 100 top cities, and Tier III as those within 55 miles of the top 100 cities.

Then, prodded by Congress, the FCC agreed to a lottery scheme to select licensees in each area. The lottery was biased to favor applicants that are at least 50 percent owned by a minority group or who own no other mass media outlets. The first lottery, of a monthly series to continue until all markets are accommodated, was held in September 1983 and decided licensees for Bowesmont, North Dakota, Milton, North Dakota, Casper, Wyoming,

Land O' Lakes, Wisconsin, Presque Isle, Maine, Steamboat Springs, Colorado, and Chillicothe, Missouri.

Technically, LPTV is the same as normal television except for the low power levels of 10 watts VHF (100 watts in clearer areas) and 1000 watts UHF. A channel unusable for full-powered TV because of another full-powered TV transmitter in the same channel in a city 150 miles away may be usable for a low-powered transmitter.

For STV operation, LPTV stations were liberated from the FCC restriction on full-powered STV to markets where there are at least four other "free" stations, and there are no minimum hours of "free" programming. Decoders must be leased, not sold, to subscribers.

It remains to be seen if LPTV will provide a boost to subscription television, but the low capital costs and possibility of multiple outlets in a single market make this an intriguing possibility.

Subscription Television's Future

Subscription television, as a self-standing alternative to cable television, will have difficulty holding a general audience, even with coordination of two or three stations in a service area to provide a multiple-channel service, although it might cater to special audiences which cable often ignores. Aside from such narrowcast programming, program distributors will be uninterested in supplying an STV operation instead of a much larger cable operation in the same community.

STV may have a permanent niche in the spaces between cable systems, helping to complete the "wiring" of America with its wireless delivery to uncabled areas. Even if the broadcast outlets for existing STV program services disappear, these services, a few of which have been distributed on the main cable programming communication satellites and sold beyond STV stations, may remain viable. When and if the switched wide-band communications network is extended to homes as suggested in Chapter 12 and elsewhere in this book, STV programmers may get another chance to reach the entertainment-consuming public.

For Further Reading

[1] *The Emergence of Pay Cable Television*, vol. 2, Technology + Economics, Inc., Cambridge, MA, July 1980.
[2] "Subscriber TV market seems to be fizzling," *Wall Street Journal*, Nov. 23, 1983.
[3] "STV: A different kind of TV station," in *Channels of Communication Field Guide*, 1983.
[4] "Court allows sale of cable decoders," *The New York Times*, Aug. 9, 1980.
[5] *FCC Low Power Television (LPTV) Fact Sheet*, BC Docket 78-253, Apr. 1982.
[6] J. Traub, "Low power TV: Broadcasting in a minor key," in *Channels of Communication Field Guide*, 1983.
[7] "FCC swamped with applications for new low-power TV stations," *Wall Street Journal*,

Oct. 30, 1981.

[8]   ''TV licensing with a lottery,'' *The New York Times*, Sept. 8, 1983.

[9]   ''Hanging the crepe for STV,'' *Broadcasting*, Oct. 15, 1984.

[10] D.M. Rice, M. Botein, and E.B. Samuels, *Development and Regulation of New Communications Technologies*, Communication Media Center, New York Law School, 1980, pp. 12–17.

[11] Telephone interview with Gordon Oppenheimer of the FCC Television Bureau, May 30, 1985.

8 Multipoint Distribution Service (MDS)

Multipoint distribution service is an "over the air" distribution medium which carries pay video programming—often a popular cable television service—in a large number of metropolitan areas. It had, at the end of 1984, a total of about 440,000 subscribers nationwide, down from a high of about 750,000, paying an average of about $16.50 per month. This loss in subscribers could be attributed to the cabling of areas previously serviced only by MDS and STV operations.

However, the FCC in July 1983 opened up eight new microwave channels, previously dedicated to the instructional fixed television service (IFTS), and changed the IFTS rules to allow educational institutions to lease their excess IFTS capacity. This stimulated a huge interest in multichannel MDS (MMDS), and a flood of almost 17,000 applications for licenses. Multichannel transmitting and receiving equipment would make possible a wireless programming service which subscribers might find comparable to cable in choice and variety. The FCC's motivation in doing this, following a successful six-month technical trial by Channel View, Inc. in Salt Lake City, was to make CATV-type programming available to millions of households not yet wired for cable.

The FCC has resorted to a lottery to select licensees, city by city, a process which may take years to complete, during which time more single-channel MDS services may go out of business. The FCC will issue two licenses in each market. MMDS may wind up as a viable service even in markets where cable is well entrenched, although it is unlikely to ever acquire a subscription base comparable to cable's. Nevertheless, as Kevin Kelley of the FCC stated in 1983, "MDS is competitive to other video services such as CATV. It's a lower cost option and definitely in the public interest" [2].

MDS (Fig. 8.1) does not, like STV, operate in standard television channels, but instead occupies special microwave frequency bands. Until the new allocations, the MDS band extended from 2150 to 2162 MHz only. This is enough for two color television channels, although only channel 1, from 2150 to 2156 MHz, is ordinarily used.

MMDS is transmitted in the four "E" channels and four "F" channels, between 2596 and 2644 MHz (as illustrated in Fig. 8.2), previously belonging to the IFTS. This assumes they are available. IFTS operators,

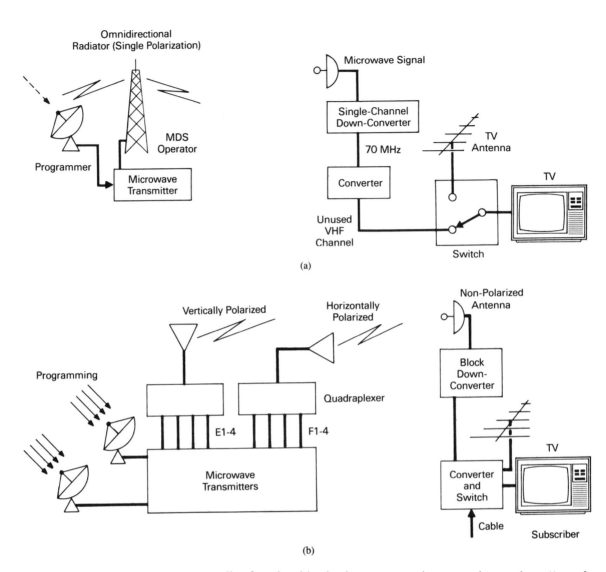

Fig. 8.1: Delivery system for microwave distribution service. For the multichannel service, the block down-converter may be equipped with a receiver capable of separating vertically from horizontally polarized signals, as well as different frequency channels, from one another. The converter may be, and for MMDS will almost certainly be, addressable and contain a descrambler. (a) Single-channel MDS. (b) Multichannel MDS.

generally educational institutions or consortiums, continue to have "grand-fathered" claims, in the cities where they are active, to those E and F channels which are already licensed to them. In the large television markets in early 1984, many of the E and F channels were taken. No E or F channels were unoccupied in Los Angeles, Sacramento, Boston, Dallas, and Houston, and only two were available in San Francisco. Moreover, MDS operators applying for use of channels adjacent to existing IFTS licensees must obtain written confirmation from those licensees that there is not likely to be adjacent channel interference, but no other grounds for IFTS-licensee objections are allowed.

Many of these existing users are expected to lease their channels, especially

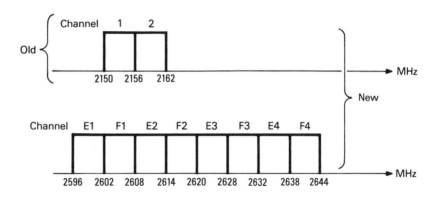

Fig. 8.2: Frequency allocations for MDS, old and new. By combining the four "E" channels into one transmission signal broadcast with a vertical polarization, and the four "F" channels into a signal broadcast with a horizontal polarization, adjacent channel interference can be minimized.

in the evening hours, to MDS operators. The FCC requires only that the educational, nonprofit IFTS licensee lease no more than 90 percent of its broadcast time to other broadcasters. Only two small MMDS systems, in Washington (Capital Connection) and Bessemer, Michigan (MI-WI TV) were in operation at the end of 1984. In the beginning of 1984, Microband Corporation, the industry's biggest, announced plans to combine its old single-channel MDS systems with IFTS channels leased from National University of the Air to provide MMDS in New York and Washington. Under the name Urbanet, Microband envisioned an expansion to more than two dozen markets over a period of five years, and in 1985 stated that Washington, New York, Milwaukee, and San Francisco were to be the locations for its first four systems. There was early criticism of the FCC decision on the grounds that it was damaging to both CATV and educational microwave services. The FCC answered that the public should have the services it wanted, and that educational licensees would be able to finance more instructional programming from its leasing revenues.

An MDS transmitter typically radiates omnidirectionally, with a maximum authorized power of 100 watts. Adequate signal reception can be expected within a 30 mile radius, although an unobstructed line-of-sight path is needed between transmitting and receiving antennas. This serious restriction eliminates part of the potential market, but the usable receiving sites will usually enjoy a good quality received signal.

A highly directive microwave antenna (Fig. 8.3) is used at receiving sites. This antenna may have a gain in signal strength, in comparison with that provided by an omnidirectional antenna, of the order of 20 decibels (dB), a factor of 100 in power. It is followed by a down-converter, pretuned to the received MDS channel or band of channels, which amplifies the signals and converts the frequencies to selected VHF channels, which can be fed directly to standard TV sets, or to CATV channels, which can be fed to CATV converters.

Transmissions can be scrambled to discourage unauthorized reception. The equipment costs (and general unavailability) prevented most single-

Fig. 8.3: *MDS receiving antenna: parabolic reflector from Conifer Corporation, shown with down-converter.*

channel operators from doing so despite the widespread unauthorized use of MDS receiving equipment. But addressable converters, with descramblers, are considered essential for MMDS, and Zenith's Z-TAC and M/A-COM's Videocipher IV were being readied in 1985 for MMDS use. Addressable converters here, as in CATV, encourage the future offering of pay-per-view entertainment. Signal theft has been a problem, and MDS operators were beginning, by the mid-1980's, to receive some support from the courts. An operator in Detroit delivering Home Box Office (HBO) won an injunction against several individuals who had intercepted its signal, and said it would send letters to many others suspected of using illegal receivers, offering to forego any claims if they would start making monthly payments.

Unlike STV, an over-the-air service which is treated as a broadcaster by the FCC, MDS is regulated as a common carrier. Licenses are good for ten years, and licensees must lease on a first come, first served basis. When it was first created in 1963, MDS was seen as a local distribution medium for closed-circuit business communications such as videoconferencing and data transmission. It was not widely used for several years, but many applications from providers of video entertainment were submitted to the FCC after it expanded its original two-channel bandwidth allocation from 10 to 12 MHz.

The first use of MDS for pay programming was in 1973, when the film "The Poseidon Adventure" was supplied to hotels in Washington, DC. Home Box Office began using MDS in 1974 for distribution to master antenna systems of apartment complexes in New York City. In 1975, Microband Corporation of America, the largest MDS carrier, entered into an agreement with Cox Cable Communications to carry pay programming in 12 large markets, and arranged program feeds via the Westar satellite. It remains true that MDS is an efficient medium for supply of programming to high-density dwellings in large cities, although it will have to quickly develop its multichannel capabilities to compete with new cable builds and expand its position as a supplier to private cable systems (see Chapter 9).

Because MDS is a common carrier and not a broadcaster, parties other than the licensees lease transmitting time, arrange for programming, and market the pay service to subscribers. The current practice is for these parties, the "programmers," to buy, install, and maintain the receiving equipment used by subscribers, and to charge a monthly subscription fee which includes rental of the equipment. The programmers purchase programming, usually the same satellite-delivered pay channels purchased by cable operators, from program distributors.

Although MDS is perceived as and often will be a competitor of cable systems, it can operate very well in a complementary mode, filling in the more distant or sparsely populated residential areas which cable operators would prefer not to wire. A number of cable operators do, in fact, have interests in MDS and are programmers for it. Some, including American Television and Communications, Group W Cable, Warner Amex Cable, Cox Cable, Colony Communications, Daniels & Associates, and Capital Cities Cable filed applications for MDS licenses.

As of mid-1984, the leading MDS carriers were those listed in Table 8.1. This set of licensees may grow with the entry of cable and private cable operators, broadcasters, publishers, satellite carriers, telephone companies, and newspaper publishers who have applied for licenses. Beside the cable operators named above, the new aspirants include: American Box Office, a private cable (SMATV) operator, and Hubbard Broadcasting, who each applied for hundreds of licenses; RCA, *The Washington Post,* and MCI, who requested several dozen licenses each; and *The New York Times,* ABC, CBS, and numerous telephone companies who asked for a handful each. These are in addition to the hundreds of licenses requested by the old MDS carriers.

The reason MDS is potentially an economical service is because the cost of setting it up is relatively low. It is far less than the cost of stringing wires for

Table 8.1: Leading MDS carriers in 1984. The number of systems includes joint ventures. (Source: LINK Resources Corp.)

CARRIER	NUMBER OF SYSTEMS
Microband Corp. of America	120
Contemporary Communications	35
Graphic Scanning Corporation	27
Telecrafter	22
Telecommunications	32
Via/Net	21
Communications Towers of Texas	6
Channel View	5

an urban cable system, even including the cost of the more expensive subscriber equipment. With economies of scale in manufacturing and customer servicing, MDS might have a significant cost advantage over cable.

The situation is particularly attractive for the MDS carrier. Cost estimates of $300,000 for a four-channel broadcasting installation and $500,000 for an eight-channel installation have been suggested, with a setup time of only one to two months. With the programmer paying the carrier a base fee of $10,000–15,000 per month for a four-channel system and $20,000–30,000 per month for an eight-channel system, plus $1/subscriber/month, the carrier can recover the capital investment quickly and have a very profitable operation.

Programmers may have a harder time. They must pay about $300 for each subscriber receiving station, and, so long as their customer base remains relatively small and scattered, they will have substantial operating inefficiencies. They are expected to charge subscribers about $25 per month for four-channel service, and $35 per month for eight-channel service. Their success will depend very much on the quality of the signals, entertainment, and service which they are able to offer.

For Further Reading

[1] "Local delivery channels: Multichannel MDS," LINK Resources Corp., New York, NY, Res. Memo., May 1984.
[2] S. Shaw, "FCC action paves way for multichannel MDS," *Private Cable*, July 1983.
[3] "FCC rule change creating a scramble over microwave TV," *The New York Times*, May 28, 1984.
[4] T. Bell, "The new television: Looking behind the tube," *IEEE Spectrum*, vol. 21, Aug. 1984.

9 Private Cable (SMATV)

Cable television, as described in the first half of this book, is a public service, franchised by and paying fees to local governments. A *private* cable system is one serving an apartment building, hotel, or institution as a closed user group, and which does not cross public property, although there is inevitably some ambiguity about this. The private cable system may be owned by the building owner or its inhabitants, or it may be owned by a separate operator who contracts with the landlord and leases services to subscribers. By one industry estimate, there are two to three million residential units in the United States which are suitable for private cable.

Most private cable systems have their own satellite receiving station headends, as shown in Fig. 9.1, and are called satellite master antenna television (SMATV) systems. They evolved naturally from the master antenna television (MATV) systems which apartment buildings and hotels have had for years, in which broadcast signals picked up by rooftop aerials were amplified and distributed to all residents. SMATV systems use headend and distribution hardware virtually identical to that of public cable systems, so that the technology discussed in Chapter 2 is also relevant to SMATV.

Because of the costs of construction and operation, an SMATV system is not practical, or at least not as economical as tying in with the franchised cable system, assuming one is available, unless there is a fairly large cluster of subscribers. A rule of thumb is that an apartment complex of 400 units or more can clearly support SMATV, and that 200 is sufficient if there is a high occupancy rate and an enthusiastic tenant population. These thresholds may move downward with the advent of direct satellite broadcasting (see Chapter 10), which can deliver satellite programming through lower cost receiving stations. There are a number of indications that DBS will actively court SMATV systems, including advertising (Fig. 9.2) by United States Communications, Inc. (USCI), the medium-powered DBS service described in Chapter 10.

Even the major pay television distributors are relenting, sometimes under legal pressure, in their reluctance to deliver programming to SMATV operations in the territories of franchised cable operators. Showtime and The Movie Channel, which have the same owner, were the first to give in. Home Box Office, after holding out for a long time, decided in early 1985 to allow affiliates (generally cable operators) to subdistribute HBO and Cinemax to

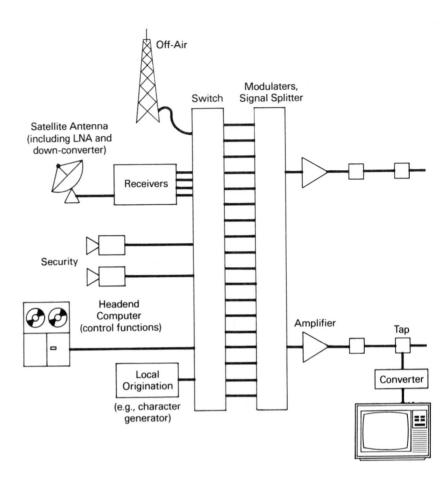

Off-Air

Satellite Antenna
(including LNA and
down-converter)

Switch

Modulaters,
Signal Splitter

Receivers

Security

Headend
Computer
(control functions)

Amplifier

Tap

Local
Origination

(e.g., character
generator)

Converter

Fig. 9.1: *Configuration of an*
SMATV private cable system. The
converters or taps may be
addressable, and two-way systems,
with data keypads at subscriber
locations, are sometimes found in
hotel systems.

SMATV operators in their franchise areas. HBO also suggested it would sell directly to some larger SMATV operators, with compensation to affiliates in the area. Many other pay and non-pay cable services, from Turner Broadcasting to The Playboy Channel, also service the SMATV industry.

Until private cable grew to the point where it could expect service from major program distributors, it relied on several smaller programming services, including SelecTV and ON TV, which willingly sold to the National Satellite Programming Cooperative, a bulk purchasing association set up by SMATV operators. The cooperative, constantly negotiating for additional programming, has also offered billing, marketing, construction, and insurance services to its members. For political action, SMATV operators work through a sister organization, the National Satellite Cable Association (NSCA).

Private cable systems in hotels may provide many additional functions beyond television programming, including pay-per-view movies, schedules of activities at the hotel, electronic billing and checkout services, fire alarms,

Fig. 9.2: Advertisement from the May 1984 issue of Private Cable magazine directed to private cable operators. The program service distributes by direct satellite broadcasting.

energy management signaling, security control, and even videoconferencing capabilities. To support these functions, hotel systems will use hardware not found in residential systems, such as computer-controlled taps and keypads in the guest rooms. Private cable systems in industrial and commercial settings may carry additional types of traffic, such as digitized voice and data communications.

Although private cable systems offer CATV-like services, they have generally provided 12 or fewer channels, and have been outclassed by new cable systems with 50 or more channels and first-string programming. As time goes by, however, SMATV is increasing its capacity, and 20- and even 60-channel systems are being offered. Sensitive to the growing attractiveness of SMATV, cable operators with newly won franchises are sometimes becoming interim SMATV operators themselves, providing programming to future cable subscribers during the construction phase of the urban cable system with the understanding that a cutover to regular cable will be made when construction is finished. Going even further, Cox Cable announced plans in 1984 to permanently install SMATV systems, free of municipal regulation and franchise fees (see the following section), in multi-unit dwellings in its New York City cable franchise, a move which was not viewed favorably by the New York State Cable Commission.

Legal Challenges

With their restriction to private user groups and to private property, SMATV systems have claimed complete exemption from local and state government regulation, including the payment of franchise fees. Franchised cable systems and local and state governments have bitterly fought SMATV, claiming it presents unfair competition which could deprive both CATV operators and cities of revenues and threaten the fundamental financial viability of CATV systems. This fight has gone on in both legal and legislative arenas.

SMATV won a major legal contest in 1983 when the Federal Communications Commission ruled favorably on the petition of Earth Satellite Communications, Inc. (ESCom), an SMATV operator in the Crescent Park apartment complex in East Orange, New Jersey, for exemption from state and local CATV regulations. The company had submitted its petition after its operations were closed down by a New Jersey court in response to a suit by Suburban Cablevision, Inc., which argued that ESCom was required to obtain a certificate of approval from the state. The petition had been opposed by New Jersey's Attorney General, the New Jersey Cable Television Association, the National Cable Television Association, and the New York State Cable Commission.

This was not the only effort by local authorities to close down an SMATV operation. In Chicago, where the city had already counted on $17 million in franchise fee prepayments, its new franchisee, Cablevision Systems, refused to sign the franchise agreement until an existing SMATV service offered by Cablecom Corporation was eliminated. The city allegedly revoked Cablecom's electrical permits in order to shut it down.

The FCC's preemption of SMATV from state and local regulation, which survived a further proceeding in a federal court, firmly established the legality of private cable as a medium regulated only by the FCC, except for local zoning and construction laws intended to protect safety and property values. "Local regulations would have been highly counterproductive in developing the programming array we now have," said FCC Mass Media Bureau Chief James McKinney. "It's important that [SMATV] not be frustrated by state and local regulations" [1].

The most bitter controversy between CATV and private cable is over the question of access to subscribers. Many private cable operations involve an exclusive agreement between operator and landlord, which CATV operators claim denies them legitimate access to potential subscribers living in multiunit complexes. Cable operators sought relief through national legislation requiring landlords to permit CATV wiring of their buildings except where private cable systems provided equivalent service, but the NSCA vigorously opposed this provision, considering it "forced access," which would allow CATV to overbuild private cable and put it out of business. Cable operators have made similar unsuccessful attempts to pass state legislation giving them access to individual renters. The Cable Act does not resolve the conflict, leaving the solution to the marketplace and the states.

The largest SMATV system in the country, with a potential for 15,000 subscribers, is in Co-op City, a huge high-rise apartment complex in the Bronx borough of New York City. It has two separate headends in order to avoid crossing the one public street in the complex. The New York State Cable Commission had decided, in early 1984, against construction of the SMATV system, but was overruled by a U.S. District Court in New York.

The Co-op City system offered, in 1984, 23 channels in its "basic" tier, including a television picture of the lobby of the building in which a subscriber lives. Seven pay packages were available, including HBO, Showtime, The Movie Channel, and WHT (see Chapter 3). Additional programming catered to the interests of the ethnic populations living in Co-op City. Addressable converters were used. The contractor who built the system, Antenna and Communications Corporation, operated smaller SMATV complexes elsewhere in the city, supplying them with 17 channels, including programming delivered by USCI's direct satellite broadcasting system.

Alternatives to Satellite Feeds

This chapter is titled "Private Cable" rather than "SMATV" because there are ways other than large satellite antennas of supplying individual private cable systems. In fact, the "headend-less" private cable system as an appendage to CATV, MDS, DBS, and other distribution networks is an operational idea which has many advantages in efficiency and economies of scale.

The association with DBS has already been suggested. The DBS operator can do most of the pay program packaging, and a low-cost receiving station is adequate, making even very small private cable systems practical. This same concept holds for supply of private cable systems by MDS and CATV operators.

A somewhat bolder concept is the wireless urban cable network (Fig. 9.3), in which a multisystem private cable operator uses a private microwave system to distribute programming to the individual private cable systems. Austin Satellite Television was an early example, supplying 24 channels to apartment complexes in the Austin metropolitan area using CARS microwave bands and AML transmission equipment (see Chapter 2). "The beauty of the microwave," according to the company's president, Chris Tyson, "is its expandability ... we simply put in a splitter and another [microwave relay] dish on the roof, and we're in business at minimum cost" [5]. In fact, this concept of using a few microwave trunks to reach concentrated groupings of communication users can be much cheaper than laying cable, and is widely used by cable operators for supplying distribution hubs, if not individual buildings. Microwave-linked networks of private cable systems clearly cross public property but appear, for the time being, to be exempt from local regulation under the FCC's 1983 ruling. Austin Satellite Television expected to have 35,000 subscribers by 1988.

The phenomenon of private cable illustrates how niches in a market can be exploited by clever entrepreneurs and, in a sympathetic regulatory

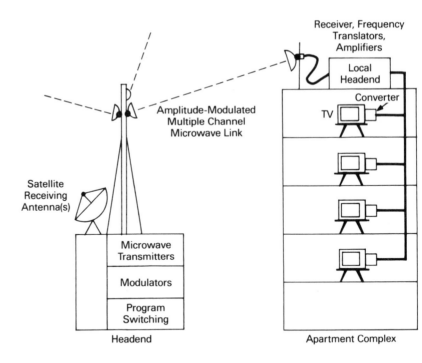

Receiver, Frequency
Translators,
Amplifiers

Local
Headend

Converter

TV

Amplitude-Modulated
Multiple Channel
Microwave Link

Satellite
Receiving
Antenna(s)

Microwave
Transmitters

Modulators

Program
Switching

Headend

Apartment Complex

Fig. 9.3: A "wireless urban cable system" created with microwave trunking links to private cable systems in apartment complexes.

environment, develop into a business category with some permanence. Private cable can fill in where CATV is not yet available, use more efficient microwave trunking systems to clustered subscriber populations, and meet the special needs, such as local security and ethnic programming, for particular tenant populations. It has the potential to provide specialized transactional, professional, and business services in hotels and institutions. It is still a small business and may never get very large, but it reduces the opportunities for CATV to become a universally dominant provider of cable services.

For Further Reading

[1] "FCC exempts private cable from state and local regulation," *Private Cable*, Dec. 1983.
[2] "Co-op City 'Goes Private,' 15,000 addressable taps," *Private Cable*, June 1984.
[3] "SMATV cleared in New York," *Broadcasting*, Mar. 12, 1984.
[4] T. Bell, "The new television: Looking behind the tube," *IEEE SPECTRUM*, Aug. 1984.
[5] "Austin Satellite TV—Model PC operation," *Private Cable*, Nov. 1983.

10 Direct Satellite Broadcasting

It has long been a dream to make communications from space, one of the most romantic and spectacular technical accomplishments of the twentieth century, available directly to ordinary people. When the Federal Communications Commission began its initial direct broadcast satellite (DBS) rulemakings in 1981, Chairman Robert E. Lee declared that "The idea of a national station or stations sitting up in the sky nearly boggles the mind" [4]. This dream is now becoming a reality through a series of massive development projects around the world. If the correct balance among the costs of satellite systems, earth stations, and programming is achieved, DBS service will soon become part of the daily lives of many millions of people.

DBS is defined by the International Telecommunications Union as a "radio communication service in which signals transmitted or retransmitted by space stations are intended for direct reception by the general public" [1]. The usual conception (Fig. 10.1) is of a receiving system dedicated to an

Drawing by C. Barsotti; © *1983 The New Yorker Magazine*

LITTLE HOUSE ON THE PRAIRIE

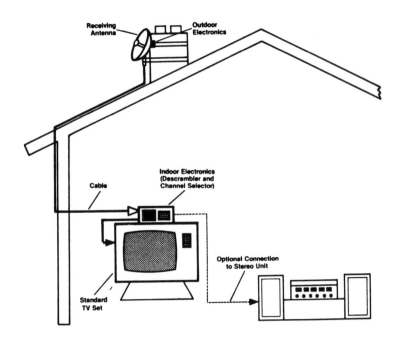

Fig. 10.1: The traditional concept of a direct satellite broadcasting receiving station. (Courtesy Satellite Television Corporation.)

Fig. 10.2: An 0.6 meter DBS antenna. (Courtesy ALCOA–NEC Communications Corporation.)

individual residence, with an antenna, low-noise amplifier, and down-converter (frequency converter) up on the roof and connected by a coaxial cable with the receiving and descrambling unit on the television set. By use of very powerful geostationary communication satellites, relatively small and low-cost receiving stations can deliver high-quality signals. A parabolic antenna reflector of less than 1 meter diameter (Fig. 10.2), and a total earth station cost of $300, have often been suggested as the criteria for mass acceptability of consumer owned or rented equipment.

A large number of North Americans—more than a million in the United States alone—have already created an unofficial DBS system through their purchases of relatively expensive large-dish receiving stations (Fig. 10.3) for the low-powered C-band satellites which carry programming intended for CATV operators, broadcast affiliates of the national TV networks, and a variety of private communications users. Most of the early buyers have been people in rural areas without access to much TV or cable service, the figurative "homes on the prairie," but many earth stations are being sold in metropolitan and suburban areas with good broadcast and cable service. This unauthorized listening into programming some of which others pay for is legal, but will be increasingly controlled by scrambling. More is said about this form of DBS later in this chapter.

Direct satellite broadcasting, in any of its forms, is not a radical departure from the CATV concept of satellite-fed cable distribution. It is only an extreme placement of the dividing line between satellite carriage and cable carriage, with the television set near the satellite antenna, and a private "cable system" within the subscriber's home.

Although intended largely for the five or six million residences outside of cabled areas, American high-powered DBS could also serve the SMATV market—the hotels, apartment buildings, and other buildings with private cable systems, especially if it offered multichannel service and superior pictures and sound. SMATV is presently limited to larger apartment or hotel complexes partly because of the costs of the headend equipment. Lower cost DBS headend equipment, together with desirable and moderately priced programming, would allow SMATV to profitably serve buildings with as few as 50 units.

There are very large differences of opinion among industry experts on the economic potential of DBS. One of the more optimistic views appeared in a 1984 report by Frost and Sullivan, a media consulting organization, which forecast 48 million North American viewers by 1994 and equipment sales of more than $3 billion. This study also forecast DBS carriage of special interest and pay-per-view programming, and mentioned the possibilities for high-definition television (see HDTV, Appendix 1).

But the costs of setting up a high-powered DBS system are immense, and, in mid-1984, two major applicants for licenses, CBS and Western Union, dropped out rather than meet an FCC deadline to begin construction. The CBS pullout was particularly damaging because CBS had been in partnership with the Comsat Corporation's Satellite Television Company (STC), the first and most ambitious applicant, and in the summer of 1984 STC moved to halve its billion dollar investment plans by falling back from four high-powered satellites to two. RCA, another prominent applicant, announced its own scaling back from four satellites with six high-powered transponders each to two satellites with 16 medium-powered transponders each, and a delay of service initiation from 1988 to 1989. United Satellite Communications, owned largely by the Prudential Insurance Company and General Instrument Company, was desperately seeking a merger with another company in 1984–85 and projecting losses totaling $119 million through 1986. A particularly pessimistic view of the industry was expressed in 1984 by financial analyst John Reidy, who stated that "It [DBS] is simply not going to happen until it gets dramatically less expensive" [5].

Assuming that technical and investment requirements, including reasonable subscriber equipment costs, are met and a consumer market really exists, there is no question that programming and marketing will be the keys to the success of DBS. The fact that DBS may not deliver as many channels as cable, although it could potentially deliver dozens, is not necessarily troublesome. "You don't have to run cable all over America to bring a narrowcast product to people who are willing to pay," said Nathaniel Kwit, Jr., president of United Satellite Communications. "Our studies have concluded that 80 percent of American television viewers watch only four channels" [2]. One of the potential operators, United States Satellite Broadcasting Company, has declared its intention to spend $168 million on programming alone in its first year of operation.

Of equal importance, perhaps, is the DBS promise of excellent signal quality. Even without high-definition television, DBS could penetrate cabled areas if it offered consistently cleaner and less distorted signals than those which are provided on many of the channels in most cable systems. This opportunity may have a limited window, since a wide-band optical fiber network extending to subscribers (see Chapter 12) would provide signals of a consistently high quality unmatched by any over-the-air medium.

It is not likely that high-powered, "official" DBS will succeed merely by filling the spaces between cable systems. Its success will almost certainly be decided by the value and viewing quality of its programming, the availability of good and inexpensive subscriber equipment, and the level of customer service which its operators can provide. Moreover, it will have to compete not only with cable and videocassettes, but also with the low- and medium-powered forms of direct satellite broadcasting described in the next section.

Low, Medium and High-Powered Services

The DBS service envisioned by the International Telecommunications Union is a high-powered service, operating in the 12 GHz K_u frequency band, with 200 watt satellite-borne relay stations (transponders), compared with the 5–15 watt, 4 GHz C-band transponders of the fixed satellite service (FSS) which feed cable systems. This is the "real" DBS, in operation only over Japan at the time of writing, which will be described in more detail in the remainder of this chapter. But the unofficial low-powered DBS

Fig. 10.3: Bob and Marge Fish of Londonderry, Vermont, with their backyard satellite dish, a steerable TVRO installation for reception from the relatively low-powered satellites carrying programming for cable systems and TV broadcast stations.

mentioned earlier, and much less pervasive medium-powered services, have already made inroads in the United States and Canada and could threaten the commercial viability of the high-powered services, which will not become available until 1986 at the earliest. In its 1984 scaling-back filing with the FCC, RCA claimed that

> It is now clear that it will be possible to serve the DBS market, as we see this market developing, with a lower power and therefore less costly satellite than originally planned. This reduction in the satellite power has for the most part been made possible by current and anticipated improvements in earth station design and performance. [6]

Despite this view, there may be a natural progression in time from a low- or medium-powered service, which can operate profitably by serving cable, broadcast, SMATV, and commercial customers, and one or two million DBS subscribers but which has relatively high equipment and servicing costs for DBS subscribers, to a high-powered service which will require three to five million subscribers to break even, but which has much lower per-DBS-subscriber costs. High-powered DBS is being made into an international standard and is regarded in many countries, if not the United States, as the legitimate heir of terrestrial broadcasting.

Low-powered DBS was, at the time of this writing, promising to become a subscription service. Some satellite program distributors, such as ESPN, Turner Broadcasting, USA Network, and MTV Networks, began in 1985 to set tariffs for private earth station owners, generally about $25 per year each. This was being done in advance of implementation of a proposed universal scrambling system. There was little motivation, aside from a possible moral one, to pay these fees, and SPACE, the TVRO industry trade association, advised its members not to pay them. In mid-1985, 40,000 to 50,000 new backyard dishes were being sold each month, but sales declined later as scrambling drew closer.

HBO and Showtime/The Movie Channel announced in 1984 their plans for scrambling, using equipment developed by the M/A-COM company. There was a strong reaction. Several bills were introduced in Congress in 1985 which would either hold off the initiation of scrambling for several years, or impose strict requirements for making descramblers and the programming itself available to home earth station owners at reasonable cost. Availability and fair pricing were the key issues, with fears that cable operators were seeking to monopolize program distribution.

HBO had, in fact, commissioned M/A-COM to develop descramblers for home earth station subscribers with a price goal of $400 or less. These descramblers would be authorized or de-authorized by data transmission through the satellite broadcast channel. But a major difficulty quickly became apparent: how could a satellite service home subscriber, able to afford only one descrambler, be able to subscribe to scrambled programming from a variety of satellite programmers, each with its own authorization apparatus and possible different scrambling system?

The solution was envisioned as a scrambling cooperative association which would operate scrambling and processing facilities and authorize TVRO decoders to receive signals. The National Cable Television Association proposed a non-profit co-op, with membership open only to cable operators, which would specify the scrambling technology, manage the subscriber data base, and provide a toll-free telephone number through which dish owners could order programming. Programmers, although offered access to facilities of the NCTA's scrambling co-op, objected to cable operator control and possible sharing of the subscriber revenues. A counter-consortium of advertiser-supported programming services was proposed by WTBS (Turner Broadcasting), ESPN, and MTV Networks. An alternative scrambling system, designed by Scientific-Atlanta, was reportedly favored by this consortium over the M/A-COM system apparently favored by the NCTA. An industry-wide agreement on a workable subscription service for home earth station owners was not in sight at the time of publication.

The home earth station market opened up with the FCC's 1979 decision to drop its prior requirement for a construction permit for any earth station. The TVRO systems, which are priced at $2000 and up, have antennas of a two or three meter diameter and are lower-cost versions of the receiving installations used by cable systems. The newer systems are steerable by remote control from the viewer's living room so that, for locations in the continental U.S., all geosynchronous C-band satellites carrying video programming destined for the U.S. can be picked out by the narrow-beam antenna, and locked onto for continuous viewing. The main difference from CATV TVRO systems is that a residential TVRO system has only one receiver, providing one channel at a time, while a TVRO installation for a cable system will have multiple receivers for simultaneous reception from many transponders on board the satellite being viewed by the antenna.

Medium-powered DBS uses satellites with transponders of 20–100 watts output power, and requires earth stations with reflector antennas of 0.75–2.5 meters in diameter, depending on location. It usually operates in the 12 GHz K_u frequency band, just as high-powered DBS does.

The receiving station is estimated to cost $500 for a station with a 1 meter antenna, considerably less than a C-band earth station but somewhat more than an earth station for a high–powered DBS service. With these parameters, a satellite can carry and power many more medium-powered transponders than high-powered transponders. This has the two large benefits of making more channels available to the subscriber and reducing the per-transponder capital investment to build the system. There are, however, questions about how good and consistent signal quality is in comparison with high-powered DBS. In fact, it may be that the effective service areas (with $500 stations) will include only part of the continental United States, that part for which the received signal level is relatively high, and the service will not be available to a large part of the American population.

There were two medium-powered systems in operation in the United States at the time of this writing, one directed primarily to residential customers and the other to commercial customers. United Satellite Communications (USCI) began broadcasting to residential customers in the vicinity of Indianapolis, Indiana, in November 1983. The service was expanded shortly thereafter to Washington, DC, Baltimore, Richmond, Cincinnati, and Harrisonburg, Virginia. Transponders were leased on the medium-powered Canadian ANIK C2 satellite.

The USCI receiving equipment was leased rather than sold. An installation fee of $300 and a monthly fee of $39.95 were asked for the initial five-channel service, which provided two movie channels, two variety channels, and a sports channel. Tandy Corporation agreed to market the service through its Radio Shack stores, and RCA Corporation to handle equipment installations.

Despite the fairly hefty subscriber fees, the company expected, when it initiated service, to reach its break-even point of one million subscribers in two or three years. However, after two years USCI had only about 9000 subscribers, was not able to raise funds for expansion, and had effectively lost the installation services of RCA's service unit.

ANIK C2 was moved in orbit to the planned location of the future Gstar satellite, the designated permanent carrier of the USCI service, so that the service could be switched from one satellite to the other without any change in the direction of receiving antennas. A later announcement suggested that five enhanced transponders on Gstar A2 would serve the eastern United States and five the western part of the country, allowing USCI Home Satellite Television to reach over 80 percent of U.S. homes and service 40 percent of this potential audience with smaller 0.75 meter antennas. It remains to be seen if these ambitious plans will be realized.

The second medium-powered service was offered by Private Satellite Network, Inc. through the SBS-1 satellite, which operates in the K_u band for business applications. The first customers were the National Education Association and the Merrill Lynch investment services company.

Among the additional consumer-oriented medium-powered services which were being planned at the time of writing was a five-channel system announced by Satellite Television Corporation (STC). STC is the Communications Satellite Corporation (Comsat) subsidiary which was the earliest proponent of a high-powered service, which it planned to begin in 1986. STC hoped to build public interest and a customer base through an interim service which would eventually be replaced by high-powered DBS.

High-powered K_u-band DBS is an internationally recognized category of satellite service, and in the early and mid-1980's was the subject of intense international negotiations and disputes. The differences were not only over frequency channels and slots in the geostationary orbit, which every country, whether or not prepared to initiate a satellite service, wanted to reserve for itself, but also over the radiation "footprints" of the powerful transponders. In Europe especially, the splashing of these footprints, with their alien programming, across national boundaries (Fig. 10.4) has been of great concern to some of the countries involved. Others, especially some of the smaller countries, have sensed an opportunity to make money and have been looking forward to large international audiences for advertiser-supported programming. Aside from intentional broadcasting to other countries, the unintentional spilling-over of interfering signals from one country to another has been a large concern and led to restrictions on satellite power emission which were not welcomed by some countries, including the United States.

Now that orbital and frequency allocations have been made by a series of

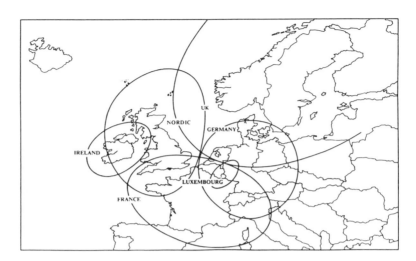

Fig. 10.4: Radiation footprints of planned European DBS satellites.

international conferences, high-powered DBS for the United States is restrained only by investment considerations. Despite the dropouts and the critical view described earlier, the remaining first-round DBS applicants for FCC licenses have shown confidence in their future, and a substantial number of second-round applicants are waiting in the wings. "I think DBS will be one of the most successful mediums that has ever been devised," said Stanley Hubbard, president of United States Satellite Broadcasting, "and it will happen much more quickly than anyone thinks" [5].

The Development of High-Powered DBS in the United States

Arthur C. Clarke, the British scientist and writer, envisioned DBS in 1945 as a system of three manned "space stations" in geosynchronous orbit (Fig. 10.5) which would broadcast to the entire world and be received with one-foot parabolic antennas. It was, of course, not practical at the time, but the idea was born. DBS as a practical service was first defined at the 1963 Extraordinary Administrative Radio Conference of the International Telecommunications Union (ITU), a U.N. agency which coordinates the telecommunications activities of all nations. In 1966, the National Aeronautics and Space Administration (NASA) requested a study from the National Academy of Sciences on "earth-oriented" applications of space satellites, and the technical panel on broadcasting which was organized by the Academy identified DBS as a technology which could improve communications to and among the different elements of American society.

Although no immediate large-scale project was initiated, small trials were undertaken in the 1970's in the United States and Canada with low- and medium-powered satellites. NASA's ATS-6 (Applied Technology) satellite broadcast to community receivers in Alaska, the Rocky Mountain states, and

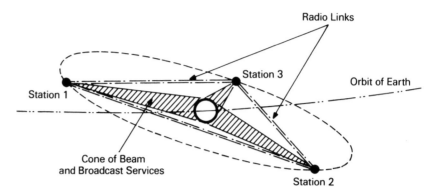

Radio Links

Station 3

Orbit of Earth

Station 1

Cone of Beam
and Broadcast Services

Station 2

Three satellite stations would ensure complete coverage of the globe.

Fig. 10.5: Arthur C. Clarke's
1945 conception of a system of
three high-powered geosynchronous
satellites for worldwide coverage.
(Courtesy Arthur C. Clarke)

the Appalachian region. The Canadian experiments with ANIK-B were the predecessors of the medium-powered DBS services described earlier.

A 1980 FCC staff report on "Policies for Regulation of Direct Broadcast Satellites" enthusiastically endorsed the DBS concept, and described the possibilities for a multiplicity of DBS systems and as many as 50 subscriber channels. The staff report recommended a minimum of regulation, and in particular no constraints on program content, prices, types of services, leasing, cross–ownership with other media, transfer of licenses, and ownership of receiving equipment. This "hands off" attitude appears to have become the long-term FCC position.

The staff report was quickly followed by an official FCC Notice of Inquiry requesting comments on U.S. requirements and objectives for a broadcasting satellite service in preparation for the 1983 Western Hemisphere Regional Administrative Radio Conference (RARC-83) of the ITU. A 1977 World Administrative Radio Conference (WARC-77) had made orbital location and frequency assignments for countries in ITU Regions 1 (Europe and Africa) and 3 (Asia and Australia), with most countries allocated four DBS channels. Monaco and Luxembourg had been given the same number of channels as Germany and France. The United States was concerned that at RARC-83, dealing with the International Telecommunication Union's Region 2 (the Americas), its large potential needs would be crowded out by assignments to small Latin American countries which would not use them for a long time, if ever.

These fears turned out to be groundless, because the United States received almost everything it wanted, including eight orbital locations (Fig. 10.6), each capable of supporting 32 channels. Technical progress in control of satellite radiation patterns allowed virtually every country to get a liberal allocation. In fact, there was a total of about 2500 different combinations of frequency, orbital location, and service area.

The U.S. was granted the alternative of four service areas, roughly

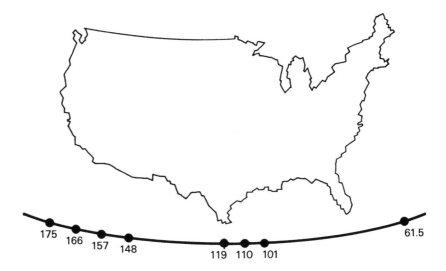

175 166 157 148 119 110 101 61.5

Fig. 10.6: Orbital assignments from RARC-83 for U.S. direct broadcast satellites, operating in the K_u frequency band. (Locations in degrees west longitude.)

corresponding to time zones, or two half-continental U.S. (half-Conus) zones, and provided for service to Alaska, Hawaii, Puerto Rico, and the Virgin Islands via spot beams. Frequency assignments (Fig. 10.7) were in the K_u frequency band, in conformance with the general policy of the ITU and the expectation of the FCC. The only setback for the United States, a minor one alluded to earlier, was that the signal power in the coverage areas was constrained to −107 decibel watts/square meter instead of the slightly higher −105 dBW/m² requested in order to support very small earth stations.

The FCC had been accepting applications for permission to build DBS systems since the fall of 1980, although only interim authorizations could be given prior to the RARC-83 decisions. The first application was received in December 1980 from Satellite Television Corporation (STC), the Comsat subsidiary introduced earlier. In 1982, eight out of thirteen applications were accepted, from CBS, Direct Broadcast Satellite Corporation (DBSC), Graphic Scanning Corporation, RCA Americom, STC, United States Satellite Broadcasting Company (USSB), Dominion Video Satellite, and Western Union. Interim rules for the licensing and operation of DBS systems were adopted in June 1982.

By late 1984, four launch permits had been issued, to STC, USSB, Dominion Satellite, and DBSC, which had proved to the satisfaction of the FCC that they were committed to going ahead with construction, a financial obligation that could total between $400 million and $1 billion. The FCC's assignment of orbital locations and channels to these companies envisioned six to twelve channels at each orbital location.

A second round of seven applicants, consisting of National Christian Network, Inc., Satellite Development Trust, Satellite Syndicated Systems, Inc., Advanced Communications Corporation, Hughes Communications

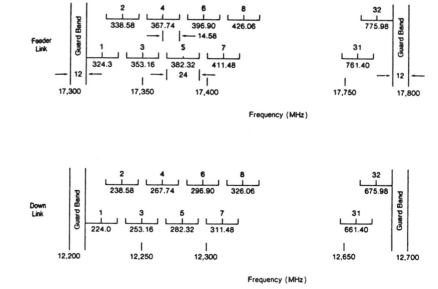

Frequency (MHz)

Frequency (MHz)

Fig. 10.7: Western Hemisphere DBS frequency assignments for 32 channels, each 24 MHz wide, in the K_u frequency band. The channels overlap, but the odd-numbered channels correspond to one polarization and the even-numbered channels to the orthogonal polarization, so that there is very little adjacent channel interference (from Rinehart [9]).

Galaxy, Inc., National Exchange, Inc., and Space Communications Service, was to be considered after disposition of all the first-round applications. RCA was put off to a possible third round because of its substantial changes in plans.

Despite the opposition of broadcasters and serious doubts about the economic viability of high-powered DBS service as now planned for the United States, it is likely that several systems will be built. There will probably be enough coordination among services to guarantee a significant number of channels in and around the major population centers of the country. If the operators are successful in capturing some of the most desirable programming, or leasing channels to the present distributors of this programming, DBS could provide serious competition to CATV.

DBS Technology

Direct broadcast satellites (Fig. 10.8) are large, heavy satellites, weighing 1000 kilograms or more, in geostationary orbit 22,300 miles above the equator. Each satellite will typically carry three to six high-powered transponders, which makes the per-channel cost of a $200 million satellite quite high. Hughes Satellite Corporation claimed to be developing a satellite with 16 high-powered transponders which would significantly reduce this cost.

DBS satellites are geostationary for the same reasons as are the low-powered satellites of the fixed satellite service. In particular, the directive antenna of a small earth station can be left fixed on a particular orbital

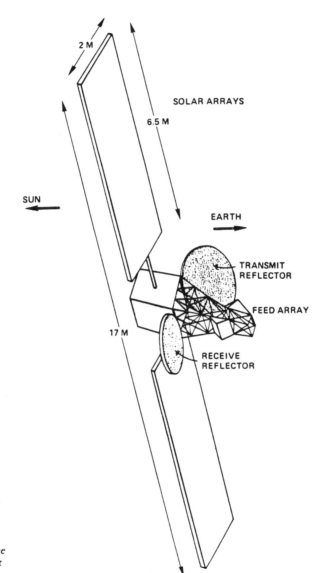

PROPOSED DBSC SATELLITE
(With Approximate Dimensions)

2 M

SOLAR ARRAYS

6.5 M

SUN

EARTH

TRANSMIT
REFLECTOR

FEED ARRAY

17 M

RECEIVE
REFLECTOR

Fig. 10.8: Design of a proposed
direct broadcast satellite, more
than 55 feet long with its solar
panels. The body of the satellite
contains six 200 watt transponders
for half CONUS coverage and
twelve 20 watt transponders for
spot beams. The solar panels
produce 4800 watts of power at the
beginning of life. (Courtesy Direct
Broadcast Satellite Corporation.)

location in order to receive programming almost without interruption from one or more satellites at that location. There are brief outages during periods of eclipse when the solar cells which power the transponders pass through the earth's shadow. By appropriate design, including orbital locations over the Pacific, the eclipses can be made to occur in the early morning hours when viewing audiences are small. Battery power can also be provided for one or two transponders, but not the full complement of the satellite. Satellites are separated by at least 9° longitude, in contrast with the 2° spacing of C-band satellites, because of the very limited resolving powers of the small earth stations.

The service can only be received by a station which has a clear view of a broadcast satellite, unobstructed by hills, buildings, or heavy foliage, and which lies within a received signal strength contour (Fig. 10.9) corresponding to the minimal signal strength requirement of the receiving station.

Signal strength could be expressed as incident power per square meter, as it was in the earlier discussion of the results of RARC-83, but it is more conventionally (and equivalently) expressed as effective isotropic radiated power (EIRP). This is, for a given received signal strength, the total power which a transponder would be radiating if it radiated equally in all directions. The 55 dBW EIRP (55 decibels above 1 watt, or 316,228 watts) which is typical for direct broadcast satellites can be provided by a transponder of only 200 watts power because that 200 watts is fed to a *directive* antenna rather than spread out in all directions. The small earth stations with antenna reflectors 30 inches across are designed for EIRP's of 55 to 60 dBW. This compares with about 35 dBW for current low-powered C-band satellites and about 45 dBW for the medium-powered satellites discussed earlier.

The direct broadcast satellites operate in two 500 MHz segments of the K_u

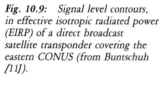

Fig. 10.9: *Signal level contours, in effective isotropic radiated power (EIRP) of a direct broadcast satellite transponder covering the eastern CONUS (from Buntschuh [11]).*

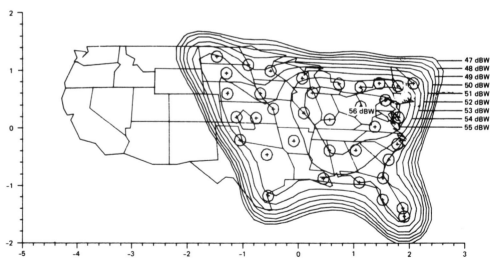

band, the segment from 17.3–17.8 GHz for the uplink from a powerful earth transmitting station to the satellite, and the segment from 12.2–12.7 GHz for the downlink from the satellite to small earth stations. The K_u band was chosen to minimize interference with terrestrial microwave systems, which are mostly in the 4–6 GHz C band, although a substantial number of existing K_u terrestrial stations were forced to give up their frequencies to DBS. Signals at K_u band frequencies are, unfortunately, attenuated by rain, so signal strengths must be high enough to give adequate received signal strength except in the heaviest rainstorms. The effect worsens as frequency increases, which is why the lower frequency segment was given to the downlink, which is more power limited than the uplink.

The 500 MHz band is broken into 32 transmission channels through use of orthogonal signal polarizations, as was already shown in Fig. 10.7. Since only six high-powered transponders are likely to be found on a single satellite, there is considerable interest among DBS operators in colocating, or clustering, satellites in a relatively few orbital locations. There is little danger of physical interference. Satellites miles apart in geostationary orbit will appear to small earth stations to be in exactly the same place. Given coordination among operators, it would be very easy to arrange a nationwide service with 15 or 20 channels.

There are many alternatives for area converage and transponder utilization in a system of two or more satellites, and Fig. 10.10 illustrates one possible arrangement. Here in the DBSC two-satellite system, in which each satellite has six high-powered transponders and twelve spot beams, the western satellite uses its six high-powered transponders to deliver a six-channel service to the entire western United States, and the eastern satellite similarly covers the entire eastern United States. The twelve spot beams on each satellite are broken into three groups of four beams, with each group delivering four channels to smaller geographical areas such as Hawaii, Puerto Rico, and metropolitan areas on the continent. This system alone can provide ten channels from one satellite to a selected metropolitan area.

The functional structure of the subscriber station is sketched in Fig. 10.11. The low-noise amplifier may be the most critical element. "Low noise" refers to the property of not contributing very much radio frequency noise from thermal effects in the amplifier itself, a problem at microwave frequencies. The standard measure of performance is the ratio G/T of gain (amplification) to noise temperature in degrees Kelvin (°K), measured from absolute zero. In a microwave amplifier, the actual thermally-induced noise in electronic components and the noise temperature have a directly proportional relationship and are equivalent. A G/T of 12 dB/°K is adequate for a DBS receiving station, and is now achievable at moderate cost with new semiconductor amplifiers.

The down-converter, also close to the amplifier, translates the signals down to a frequency band near 500 MHz which can be transmitted through a coaxial cable to the receiving unit near the subscriber's television set. After

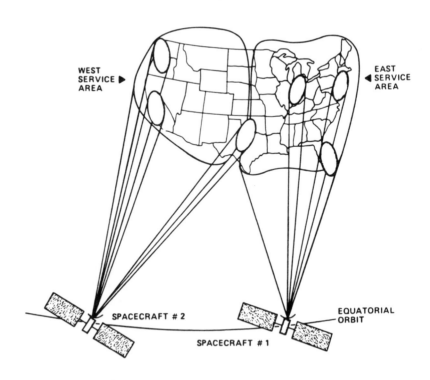

WEST
SERVICE ▶
AREA

EAST
◀ SERVICE
AREA

SPACECRAFT # 2

EQUATORIAL
ORBIT

SPACECRAFT # 1

Fig. 10.10: A suggested operational plan for the two-satellite system of the Direct Broadcast Satellite Corporation. Each half-CONUS area is served by six high-powered transponders, and four-channel spot beams serve smaller areas. (Courtesy Direct Broadcast Satellite Corporation.)

further amplification, a second down-converter with a tunable local oscillator picks out a desired channel. The FM-modulated signal is demodulated, and the resulting baseband video signal is descrambled, if necessary, as it almost certainly will be, and converted into a standard NTSC television signal for feeding to a television set.

FM modulation is used on the satellite link for its greater resistance to noise and distortion. A special video format called multiplexed analog component (see Appendix 1), not NTSC, is used as the modulation signal in order to avoid the damage which the high-frequency color subcarrier of an NTSC signal might suffer from FM noise. This system time-compresses (e.g., speeds up) scan lines and inserts color information in the freed portion of the scan time. Two or more audio channels can be digitized and sent as data during part of this time. The decoders need to have a line storage capability in order to convert to NTSC format.

Aside from the requirement for a stronger received signal, to compensate for the reduced signal-gathering capability of a small reflector antenna, the TVRO station is much like those that have been used for years in commercial installations. Adaptations are planned for the residential market, such as pre-adjusting antennas according to the zip code of the purchaser to simplify

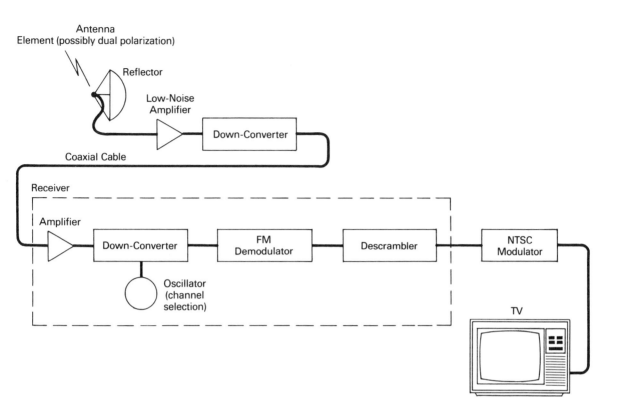

Fig. 10.11: Functional elements of a DBS earth station.

antenna aiming during the installation process, but the technologies of satellite communications are proven through many years of use.

Whatever the financial difficulties, it is hard to imagine that an idea as exciting and open to new applications as direct satellite broadcasting will not become commercially viable. Like so many other technologies, it will not exist in a pure or simple form, but in many different forms, with satellites of different transmitted powers, area coverages, and service functions. It could become both a competitor of CATV and a means of extending the reach of cable programming services and cable systems themselves.

For Further Reading

[1] ITU Regulation 84AP Spa2, quoted in "FCC staff report on policies for regulation of direct broadcast satellites," Sept. 1980.
[2] "Marketing, programming are DBS keys," *CableVision*, Apr. 18, 1983.
[3] T. Bell, "The new television: Looking behind the tube," *IEEE Spectrum*, Aug. 1984.
[4] *FCC News*, Rep. #16318, Apr. 21, 1981.
[5] "Satellite TV systems seem in doubt," *The New York Times*, July 12, 1984.
[6] "More setbacks for DBS," *CableVision*, July 23, 1984.

[7] A. C. Clarke, ''Extraterrestrial relays,'' *Wireless World*, Oct. 1945.

[8] W.G. Stallard, ''C-MAC—A high quality television service for DBS,'' *IEEE Trans. Consum. Electron.*, vol. CE-29, Aug. 1983.

[9] E. E. Reinhart, ''An introduction to the RARC '83 plan for DBS systems in the Western Hemisphere,'' *IEEE J. Select. Areas Commun.*, Jan. 1985.

[10] J. C. McKinney and G. A. Fehlner, ''The flexible domestic regulatory approach for direct broadcast satellite systems in the United States,'' *IEEE J. Select. Areas Commun.*, Jan. 1985.

[11] R. F. Buntschuh, ''First generation RCA direct broadcast satellites,'' *IEEE J. Select Areas Commun.*, Jan. 1985.

11 Videocassette Recorders

It comes as a surprise in this age of electronic communications that retail stores, rather than wires or radiated signals, could become the leading distribution channel for video entertainment. The rental and sale of videocassette tapes has become a major business. In mid-1985, 23 percent of all U.S. television households owned videocassette recorders (VCR's), and 11.5 million units were expected to be sold that year. The average price of a VCR fell from $771 in 1980 to around $400 in 1985, with the expectation of even lower prices when Korean-made sets entered the market. It was noted that the history of sales of VCR's was paralleling that of color television during its takeoff years.

A Supreme Court decision in January of 1984, in favor of the defendant, the Sony Corporation, and against the plaintiffs, Walt Disney Productions and Universal Pictures, upheld the legality of home taping of broadcast copyrighted material and gave a large boost to the already dramatic growth in VCR popularity. With forecasts of as many as 37 million VCR's in U.S. homes in 1987 (Table 11.1), prerecorded tapes could eclipse CATV as the principal vehicle for home distribution of motion pictures, and even begin to appeal to studios as a first-run outlet to be used in parallel with movie theaters. Already in 1984, a contemporary Hollywood movie was earning 13 percent of its revenues from sales of cassettes and videodisks. "Home video," in the words of Richard Snyder, chairman of the Simon and Schuster book publishing company, "is going to be the next major mass medium" [6]. (For numbered references, see "For Further Reading" at the end of this chapter.)

Beyond its value for mass marketing of video entertainment to the VCR-owning public, the VCR medium could have a considerable impact on the packaging and dissemination of ideas. Monroe Price, Dean of the Benjamim N. Cardozo School of Law noted that the VCR medium eliminates the "gatekeepers," [4] corporate in the United States and often governmental elsewhere, who regulate and define the flow of video information to theaters, over-the-air audiences, and CATV subscribers. VCR's are being used in the U.S.S.R. for distribution of Western movies and other forbidden taped materials, and the cultural restrictions of some fundamentalist Moslem countries are being similarly undermined. In the United States, the long-sought "narrowcasting" of material of interest to small audiences may be sooner realized on tape than on cable. Individual production is being

Table 11.1: Growth of the VCR market: sales of machines, blank tape cassettes and prerecorded cassettes.*

Year	VCR's Sold	Prerecorded Cassettes Sold	Blank Cassettes Sold	VCR Households	Pay Cable Households
1980	805,000				7,800,000
1981	1,361,000				11,800,000
1982	2,035,000	8,800,000	24,000,000	4,500,000	14,000,000
1983	4,150,000	9,500,000	57,000,000	9,000,000	21,500,000
1984	7,000,000	15,300,000	77,000,000	15,100,000	24,000,000
1985	11,500,000	22,400,000	—	26,500,000	26,000,000
1986	6,000,000	27,300,000	—	31,500,000	30,000,000
1987	6,000,000	32,000,000	—	37,500,000	33,150,000

*Sources: Electronic Industries Assn., Yankee Group, CableVision, The New York Times, and LINK Resources Corporation. 1984 and later figures are estimates.

encouraged by a new generation of consumer-oriented video cameras and recorders.

There are several good reasons for the rise of the VCR medium. Foremost among these is the technical progress, described later in this chapter, which produced inexpensive and very long playing tape formats and compact and economical machines on which to play them. The fact that a number of different systems were offered to the home market and that two incompatible cassette formats, Betamax and VHS, survived the marketplace competition did not diminished public enthusiasm, although it did sometimes caused confusion. The very large VCR sales of the mid-1980's reflect a growing consumer comfort with the machines, as well as lower prices. Manufacturers felt confident enough of public understanding and acceptance to introduce a third format, the miniaturized 8 millimeter cassette, in 1984.

Given this equipment, consumers were first motivated by the opportunity for personal taping, usually for "time translation" to allow viewing of broadcast programs at more convenient times, but sometimes for building up film libraries for personal use or for lending to friends. A 1984 Nielsen VCR usage report found that 85 percent of tape playbacks were of tapes made from television broadcasts, and only 15 percent from prerecorded tapes. Broadcast taping is the practice which provoked the motion picture producers into their unsuccessful copyright infringement suit against the Sony Corporation. It may become less of a threat to the producers as they become more skilled at acquiring a large share of revenues from the sale and rental of prerecorded cassettes, but it is still a bone of contention, as described later. Cable operators hope, a little unrealistically, that time translation for personal viewing, which helps rather than detracts from pay subscriptions, will continue to be the major use of VCR's.

The next surge in sales came from the growing availability and choice of

"Lester, shouldn't we be upgrading something?"

prerecorded materials. The combination of a large selection of prerecorded films and other materials with an unbundled pay-per-view system of paying only for what you want has found a great deal of consumer acceptance. Daily rental fees are usually three to five dollars, but can be two dollars or even less. With the convenience, in mid-1985, of about 20,000 retail outlets, most carrying more than 1000 titles, and some as many as 7000 (Fig. 11.2), this choice has become real for consumers almost everywhere in the country. No other medium can offer anything approaching this large a selection.

Part of the popularity of prerecorded videocassettes, and extremely

Fig. 11.1: Betamax and VHS
videocassette recorders. (Courtesy
Sony Corporation of America and
JVC Corporation of America.)

significant as a competitive factor, is that movies usually become available on videocassettes only three to six months after theatrical release, which is four to five months before they are broadcast on pay television. This is, of course, a conscious policy of the motion picture studios, who know that pay-per-view, whether at a box office or a tape rental store, is the sales system which brings in the most revenue from the early showings.

Videocassettes have apparently become CATV's primary competitor for the home video enthusiast's dollar. Once the VCR machine is paid for, the running cost for movies, at a reasonable viewing level, is comparable to subscription to a pay cable channel. Tee Yakura, a representative of Hitachi, a VCR manufacturer, declared in 1984 that "Cable can't touch us. Cable is too far behind and they don't have the marketing structure" [1]. It remains to be seen how cable will respond to this serious challenge.

VCR Technology

Videotape recording, like audio tape recording, stores information by magnetizing minute areas of the magnetically sensitive material coated on the tape. This is done in accord with an electrical signal which controls the strong, but very localized, magnetic field generated by the recording head to which it is supplied. The information content (and consequently the

Fig. 11.2: Consumers are attracted by the huge selection of movies at some videocassette retail stores. (Courtesy Cine Club Video.)

bandwidth) of a video signal is, however, so much larger than that of even a high-fidelity stereo music signal that much different recording techniques have to be used. The relative speed of the tape and recording head has to be very high to capture the higher frequency components of the video signal. It is not enough to simply run a tape past a fixed recording head. The story of modern VCR's is essentially that of developing recording techniques which paint very narrow tracks of information on a tape at a very much higher speed than that at which the tape is moving through the recording mechanism.

Videotape recording was developed in the 1950's. Bing Crosby Enterprises first demonstrated black and white recording in late 1951, using fixed recording heads. Ten parallel tracks, plus two additional tracks for synchronization and sound, were recorded on a 1 inch tape, which ran at 100 inches per second. The video bandwidth was less than half of the standard 4 MHz, so that quality was not very good. In 1953, RCA showed a videotape recorder using 1/4 inch tape moving at 360 inches per second, with four tracks: one for the black and white signal, and three others for the red, blue, and green color signals. The bandwidth was much improved on this machine, and picture quality was good. But because of the high tape speed, a 17 inch reel of tape was good for only four minutes of recording.

A great step forward was made with helical scanning (Fig. 11.3), introduced by the Ampex Corporation in 1956 and used, with variations, in today's consumer VCR's. This system scanned in four parallel tracks diagonal to the direction of tape motion. The recording heads were mounted on a rapidly rotating drum around which the tape moved in a helical rather than a circular way, i.e., the tape was guided downward as it circled the drum. The narrow diagonal tracks recorded on the tape were swept at a rate (inches per second) closer to the circumferential velocity of the drum than the relatively slow forward speed of the tape. The forward tape speed was reduced to 15 inches per second, yielding 64 minutes of recording time on a 12 inch reel of 0.001 inch thick tape. The video helical-

Fig. 11.3: Comparison of longitudinal and helical tape recording. (a) Longitudinal recording: the tape moves past a stationary recording head, producing a magnetic track in the direction of tape travel. (b) Helical recording: the tape engages a drum which carries the recording heads and is rotating rapidly, perhaps at 1800 revolutions per second, in a helical fashion, producing magnetic tracks diagonally across the tape. Audio and control tracks are recorded linearly along the edges of the tape. (Source: Encyclopedia Americana, 1976 edition.)

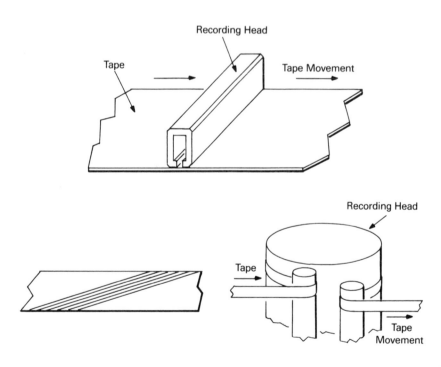

scanning recorders, made commercially available in 1959, were at first used only for studio recording.

Helical-scanning recorders were introduced in a bigger way by the Sony Corporation in the late 1960's with its "U-Matic" recording format, using 3/4 inch tape, now on a cassette instead of a reel. The machine itself opened the cassette and automatically threaded the tape. Tape speed was reduced to 7.5 inches per second, and the machine used a simpler two head configuration. Manufactured by Sony, Panasonic, JVC, and others, the U-Matic machines and tapes enjoyed widespread use in the broadcast industry, but did not sell well to the general public. Its limitations included a relatively high $30 price for a tape cassette limited to 60 minutes of recording time.

Sony continued to work on an acceptable consumer recorder, and by the mid-1970's had produced its first Betamax recorders. In addition to introducing smaller cassettes with 1/2 inch tape, the Betamax used a denser recording technique, called azimuth helical scanning, which allows tracks to be put next to each other without the guardband used in earlier systems. Sophisticated mechanical and electrical design, in particular small, opposite deflections in azimuth angle (with respect to the drum axis) of the recording heads mounted on opposite sides of the drum, permits playback without the mutual interference between adjacent tracks which one would expect. Very small recording head gaps, about 0.6 micrometer for Betamax machines, were also necessary to avoid muddying the recording of the wide-bandwidth video signal. With the recording head rotating at 1800 revolutions per second, the relative tape speed was 6.9 meters per second (273 inches per second) while the actual tape speed was less than 2 centimeters per second. Long recording times became possible, exemplified by the five-hour Beta III introduced in 1979.

Most of the recorders sold to consumers use two recording heads instead of four. There is no significant difference at the higher recording speeds, but the greater continuity of head-to-tape contact of the four-head machines provides a better quality picture at slow speeds and for freeze-frame.

The VHS (Video Home System) recording format, developed by the JVC company in Japan, was introduced to the U.S. market less than two years after Betamax. It uses an azimuth helical scanning technique and recording format very similar to that of Betamax, although the effective relative tape speed (at standard play) is 5.8 meters per second (228 inches per second) instead of Beta's 6.9 meters per second, the head gap is somewhat smaller, and there are other, relatively minor differences between the two recording systems. The most prominent differences between the two formats are in the cassette construction and tape drive mechanisms. All in all, the two systems are completely incompatible, so that a Betamax videocassette can be played only on a Betamax machine and a VHS cassette only on a VHS machine.

Aside from external dimensions (the VHS cassette is somewhat smaller), the major differences in the cassette-tape drive mechanisms are as follows (see Fig. 11.4):

1) A Betamax tape is automatically threaded in the tape transport when the cassette is put in the machine, and kept there until the cassette is removed. Considerably more tape is threaded than for VHS, increasing loading/eject time but allowing faster fast forward because of reduced tape strain. The VHS tape can be retracted into the cartridge for rewind, with the advantage of less head and tape wear.

2) For end of tape sensing, the Betamax tape has metallic leaders and trailers which are sensed by electromagnetic circuits, while the VHS tape has clear leaders and trailers through which photocells can detect the light from a small "sense lamp." The VHS cassette has holes that align with that lamp and with sensors in a horizontal plane. In older VHS recorders, the incandescent sense lamp would sometimes burn out and (temporarily) disable the recorder, but newer models have long-lasting light-emitting diodes.

Although Sony claims superior picture quality for Betamax, and has been a technical innovator, there is no clear, overall advantage to either system. The present four-to-one sales lead of VHS in the United States is due entirely to marketing. Many other VCR features, such as automatic recording at preprogrammed times, are unrelated to the recording system differences. Further technical advances, such as high-quality stereo sound, have already been introduced, and could influence future consumer preferences.

Passing reference was made at the beginning of this chapter to the new 8 millimeter format for personal videotape cameras, which was in part a challenge by American manufacturers such as Eastman Kodak, General Electric, and the Polaroid Corporation, but became a major thrust for Sony. The videotape recorder can easily be built into the camera (Fig. 11.5), creating a "camcorder." As Akio Morita, chairman of the Sony Corporation, put it, "the 8 millimeter video recorder has the definite advantage of being small in size and light in weight, so we're convinced it is the next generation VCR" [7], and this conviction has been backed up by investment and manufacturing by Sony and others. The 8 millimeter cassettes run for about 90 minutes.

But the Japanese manufacturers are producing smaller "VHS-C" cassettes also, retaining the 1/2 inch tape width while reducing the package dimensions, other than thickness, to about the same as an audio cassette. The playing time is 20 minutes. The VHS-C cassettes can, with the aid of an adapter, be played on regular VHS VCR's. Reductions in size and weight of standard-format recorders appear to be possible using four-head mechanisms with smaller drums.

The electronic processing for videocassette recording is sketched in Fig. 11.6 (see pg. 196). In order to meet bandwidth and distortion suppression requirements of the magnetic tape recording medium, the signal is not

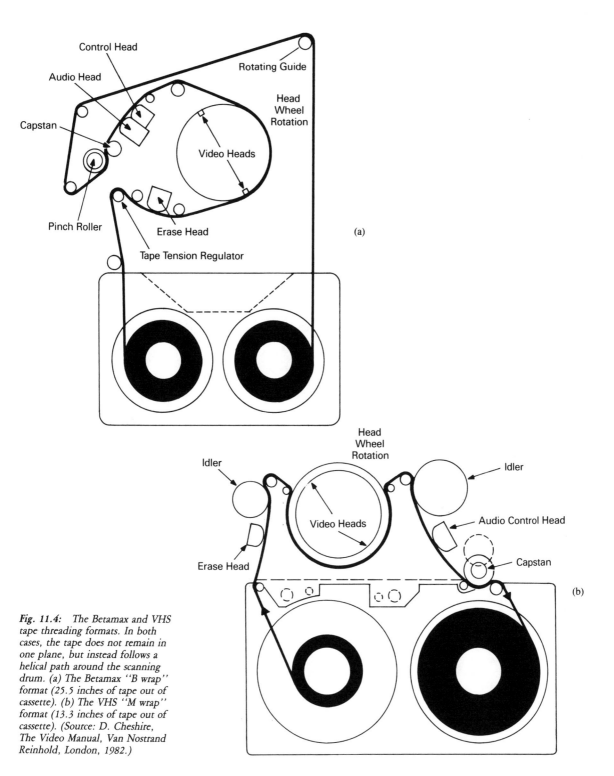

Control Head

Audio Head

Rotating Guide

Head
Wheel
Rotation

Capstan

Video Heads

Pinch Roller

Erase Head

Tape Tension Regulator

(a)

Head
Wheel
Rotation

Idler

Idler

Video Heads

Audio Control Head

Capstan

Erase Head

(b)

Fig. 11.4: *The Betamax and VHS tape threading formats. In both cases, the tape does not remain in one plane, but instead follows a helical path around the scanning drum. (a) The Betamax ''B wrap'' format (25.5 inches of tape out of cassette). (b) The VHS ''M wrap'' format (13.3 inches of tape out of cassette). (Source: D. Cheshire, The Video Manual, Van Nostrand Reinhold, London, 1982.)*

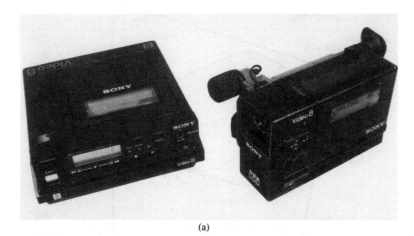

(a)

(b)

Fig. 11.5: Combination video cameras and recorders using smaller cassettes. (a) Sony ''Mini-8'' camrecorder and portable recorder/player. (b) JVC's GR-C1U camcorder using VHS-C cassettes. (Courtesy Sony Corporation of America and JVC Corporation of America.)

recorded in NTSC composite video format. Instead, the luminance signal (see Appendix 1) is FM modulated and the *I* and *Q* chrominance signals are moved down in frequency to a band around 688 kHz. The reasons for doing this, deriving from the peculiarities of magnetic recording, include the FM-modulated luminance signal serving as a bias signal for recording the chrominance signal, and good separation in frequency of the two signals. The luminance bandwidth may be less than 3 MHz, considerably less than the 4.2 MHz of the NTSC standard, and picture quality does not compare favorably with a good quality broadcast signal.

These recording procedures are reversed in playback. The recreated composite video signal may be further modulated onto a radio frequency

carrier so that it can be received by an ordinary television set, or used directly as input to a video monitor.

VCR "Software": Growth and Controversy

Motion pictures constitute the bulk of tape recorded materials, and the motion picture industry has perceived VCR's as both a threat and an opportunity. For some years, it was mainly a threat. Producers still felt, when this book went to press, that home taping of copyrighted materials should be subject to direct or indirect royalty taxes. This was a second-line defense after the defeat, in the U.S. Supreme Court, of the studios' "Betamax" suit to prohibit home taping of broadcast movies.

The main proposal was for a royalty tax, added to the retail price of VCR's and blank cassettes, which would be collected by the U.S. government and turned over to the film producers. The Motion Picture Association of America (MPAA) favored additional legislation which would give film producers more control over the rental of prerecorded cassettes, which has been at the discretion of retailers without any requirement for payment of royalties to the original producers. The 1976 copyright law contains a "first sale doctrine" which allows any buyer of a copyrighted work to rent that material without restriction.

Opponents of the proposed tax claimed that the movie producers already get most of the large revenues from the home video market, estimated at one billion dollars in 1983, and that the tax would pay the studios repeatedly for the same thing. There were also serious objections to treading

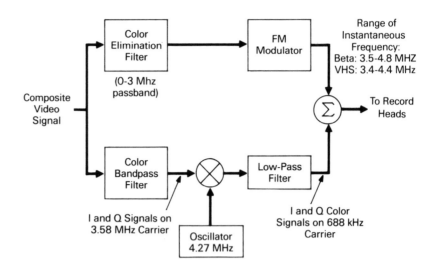

| | | Range of Instantaneous Frequency: Beta: 3.5-4.8 MHZ VHS: 3.4-4.4 MHz |

Fig. 11.6: *Signal processing for VCR recording.*

on personal freedom in the home. Senator Dennis DeConcini of Arizona stated in 1984 that

> Congress never intended to deprive the individual of his or her right to tape signals legally entering their living room for replay at a more convenient time. Instead, it is this Senator's view that once a signal crosses the threshold of an individual's home, all copyright obligations cease, unless the signal is captured for commercial purposes. [5]

However this effort by the producers for more control of personal taping and of cassette rentals turns out, they are going to continue to develop the sale of videocassettes as a significant distribution channel. It was estimated that one and one-half million cassettes of "Beverly Hills Cop" were sold in 1985. All the movie companies together sold 20 million cassettes in 1984. According to *Billboard Magazine*, 90 percent of all films grossing $25 million or more in the last 50 years, and 60 percent of all films grossing $4 million or more, have been released on tape.

Although feature films constituted 80 percent of cassette sales in 1984, a market is developing for other materials. Ten percent of the tapes sold were music video, and 5 percent were children's programming. Specialized material of considerable interest to small audiences, covering topics such as science and art, is beginning to become available. Jane Fonda's *Workout* tape sold over 275,000 copies from its release in 1982 to 1984, but this was a major exception.

With tens of millions of VCR's in use, it has become easier to reach the roughly 10,000 unit threshold of profitability for cassette release. Cassettes will be increasingly relied upon by film and non-entertainment-video producers for pay-per-view first- or second-run distribution. If video and sound quality can keep up with consumer expectations, and cassettes, blank and prerecorded, do not become unreasonably expensive because of producer and distributor efforts to squeeze income out of them, videocassettes will probably become the leading distribution medium. CATV is responding with pay-per-view services, but only a very high capacity wired distribution system, such as the fiber optic network described in the next chapter, could hope to match the price and choice standards set by videocassettes.

For Further Reading

[1] "'VCR craze' talk of CES in Chi Town," *CableVision*, June 25, 1984.
[2] "Nonentertainment video and home computer software," LINK Resources Corp., New York, NY, New Electronic Media Program, 1984.
[3] B. Pasternack, *Video Cassette Recorders—Buying, Using and Maintaining.* Blue Ridge Summit, PA: Tab Books, 1983.
[4] M. E. Price, "The videotape revolution," *Wall Street Journal*, Mar. 5, 1984.
[5] "Tax sought on cassettes and home recorders," *The New York Times*, July 2, 1984.
[6] "The video revolution," *Time Magazine*, Aug. 6, 1984.
[7] *Billboard*, June 30, 1984.
[8] J. D. Lenk, *Complete Guide to Videocassette Recorder Operation and Servicing*, Englewood Cliffs, NJ: Prentice-Hall, 1983.
[9] D. Cheshire, *The Video Manual.* London: Van Nostrand Reinhold, 1982.

12 The Telephone Network

The public communications network is run in the United States by a collection of telephone companies and carriers employing millions of people and operating billions of dollars worth of sophisticated transmission, switching, and information processing equipment. But despite its size and daily use by almost everyone, its operations are invisible to most users. The average subscriber expects fast, automatic, and clear connections to telephones nearby and thousands of miles away as part of the natural order of things, and is totally unaware of the complexity and potential of the systems providing this basic service. This potential includes the full range of video services delivered by CATV and other wide-band media.

The multiple webs of switching, processing, and transmission facilities which exist today are evolving into a more flexible, much higher capacity, and more functionally integrated communications infrastructure. Future subscribers will be offered a wide range of personalized, multimedia, user-defined services. It is doubtful that CATV, or any other alternative public communications network, can achieve anything approaching the capabilities for performance, innovation, and major construction which exist today in the U.S. telephone industry. This is despite the deregulatory upheavals of recent years, which did away with the American Telephone and Telegraph (AT&T) monopoly and scattered responsibility for communications innovation and implementation among a large and varied collection of carriers and equipment manufacturers.

The telephone network already dominates voice and data transmission. For delivery of video programming, a future all-digital network reaching subscribers with optical fiber instead of twisted pairs of copper wire could offer far superior service, as suggested several times in this book and described later in this chapter. There is a good chance that this switched optical fiber network will in fact be built, and that present-day cable operators, like other communications services providers, will become program distributors on this network rather than through their own coaxial cable systems.

Much has been written about the changes in the telephone industry's structure, technology, and services. Many new possibilities, as well as problems, were opened by deregulation and the historic 1984 divestiture by AT&T of its operating telephone companies. AT&T and the seven independent regional holding companies have entered into new competi-

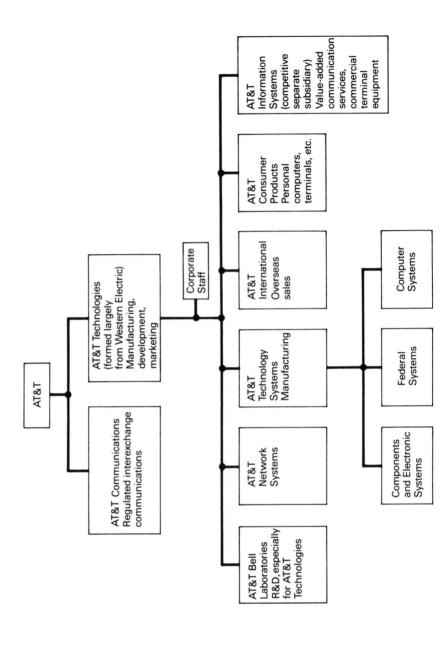

AT&T

AT&T Communications
Regulated interexchange communications

AT&T Technologies (formed largely from Western Electric) Manufacturing, development, marketing

Corporate Staff

AT&T Bell Laboratories R&D, especially for AT&T Technologies

AT&T Network Systems

AT&T Technology Systems Manufacturing

AT&T International Overseas sales

AT&T Consumer Products Personal computers, terminals, etc.

AT&T Information Systems (competitive separate subsidiary) Value-added communication services, commercial terminal equipment

Components and Electronic Systems

Federal Systems

Computer Systems

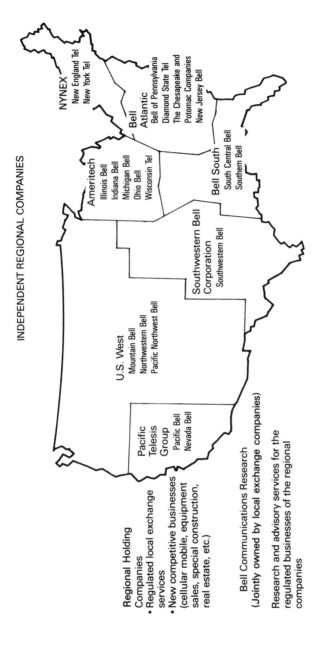

INDEPENDENT REGIONAL COMPANIES

NYNEX
New England Tel
New York Tel

Bell Atlantic
Bell of Pennsylvania
Diamond State Tel
The Chesapeake and
Potomac Companies
New Jersey Bell

Ameritech
Illinois Bell
Indiana Bell
Michigan Bell
Ohio Bell
Wisconsin Tel

Bell South
South Central Bell
Southern Bell

Southwestern Bell
Corporation
Southwestern Bell

U.S. West
Mountain Bell
Northwestern Bell
Pacific Northwest Bell

Pacific
Telesis
Group
Pacific Bell
Nevada Bell

Regional Holding
Companies
• Regulated local exchange
  services
• New competitive businesses
  (cellular mobile, equipment
  sales, special construction,
  real estate, etc.)

Bell Communications Research
(Jointly owned by local exchange companies)

Research and advisory services for the
regulated businesses of the regional
companies

Fig. 12.1: (above) *Organizations born in 1984 out of the Bell System (below) Independent regional companies.*

tive businesses, including computer manufacturing, cellular mobile telephony, joint ventures with foreign companies, and a variety of innovative telecommunications services, and are pursuing aggressive marketing and product development strategies which would have been unthinkable in the pre-divestiture days. Fig. 12.1, defining the organizations created from the old Bell System, suggests some of the new directions for the telephone industry, which has, of course, many participants in addition to those which emerged from divestiture.

Telephone companies, such as the Bell companies, others owned by GTE and Continental Telephone, and a number of small "independents" have near monopolies in their operational subregions, called local access transport areas (LATA's, Fig. 12.2). Still, they are facing bypass competition from CATV operators, microwave radio services, and others who concentrate on routing bulk business traffic around them. The issue of whether bypass is "economical," offering lower prices because of lower operating costs, or "uneconomical," offering lower prices because of regulatory constraints on tariffs, is sometimes hotly debated.

Between LATA's, traffic is carried by interexchange carriers such as AT&T Communications, MCI, and GTE Sprint. Somewhat separate from these are the specialized data networks such as GTE Telenet, Tymnet, and networks operated by AT&T and IBM. This new environment in which at least three carriers are involved in each inter-LATA long distance connection is illustrated in Fig. 12.3.

In this brief chapter, the goal is only to show what the telephone industry

Fig. 12.2: Local access transport areas (LATA's) operated by NYNEX in New York State. Some LATA's extend into neighboring states. Traffic within LATA's is, in general, handled by local exchange carriers, and between LATA's by interexchange carriers.

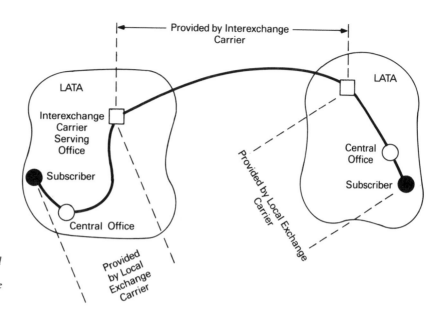

Fig. 12.3: *The minimum combination of local exchange and interexchange carriers needed to complete an inter-LATA telephone circuit.*

is doing or could do in those service areas where the cable industry sees its future. This category *could* include individual subscriber voice services, the bread and butter of the telephone industry, in which cable has shown a tentative interest, but will be limited to data and video services for which a long-term competition between the two industries is more likely. "Data" should be understood as an umbrella term taking in everything from signaling a few numbers from a keypad to bulk transfer of digitized voice and computer data among locations of large business users.

Data Communications

Most of the present and potential competition between the telephone network and CATV is in data transmission, including consumer-oriented services such as videotex, and bulk commercial data communications such as point to point links operating at data rates of 56 kilobits/second and above. The telephone network has a head start and significant strengths in both areas, and is taking steps to maintain its lead. The Cable Act, as explained in Chapter 5, effectively, but not necessarily permanently, placed data communications on CATV systems under the same state regulatory umbrella as are data services in the telephone network.

Interactive data communications is used for videotex, personal computer communications, point of sale transactions, and a wide variety of other personal and business applications. It has for many years been offered by the telephone network through dialed and private telephone lines, as illustrated in Fig. 12.4. A private line is a connection through the network which is

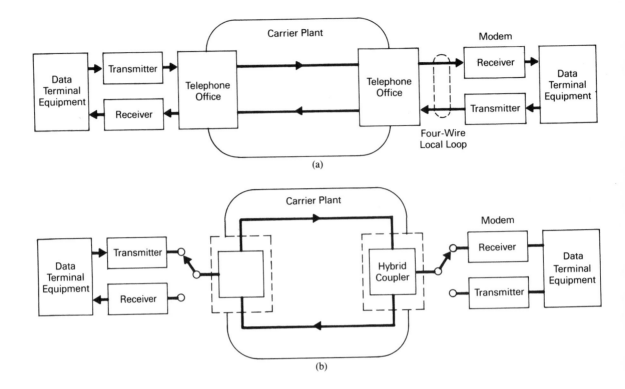

(a)

(b)

Fig. 12.4: Telephone circuits for data communication. (a) Private line connection, typically four-wire for full duplex (simultaneous two-way) data communications. (b) Dialed connection, shown operating half duplex (one way at a time). Full duplex operation is possible, usually at lower rates, through bandsplitting or echo cancellation techniques. The hybrid couplers match the conventional twisted pair subscriber line to carrier systems which have a separate channel in each direction.

dedicated to a particular user, and is left in place at all times for immediate use. Its advantages include instant call setup and, for analog telephone lines, circuit improvements, called "conditioning," which reduce transmission impairments. Most private lines are "voice grade," i.e., with a transmission capacity comparable to a voice circuit. They are also available with larger capacities, and in digital versions at data rates into the megabits per second. The private line is too expensive for occasional use. Almost all consumer applications, and many low-density business applications, are practical only with dialed telephone channels or the shared data networks described later.

Although voice-grade telephone channels were designed for voice communications and have many limitations for data service, including a relatively narrow bandwidth of about 3000 Hz inconveniently running from 300 Hz to 3300 Hz (Fig. 12.5), ways have have been found to reliably send data through them at rates as high as 14,400 bits per second. The instruments designed for this purpose are called *modems*, from *modulator-demodulator*. They have the main functions of converting the data stream into a modulated waveform which can pass through a telephone channel, compensating for channel distortions, and converting the received signal back into a data stream (Fig. 12.6). The lowest rate and least expensive modems, still common for personal computers at the time of writing, use a

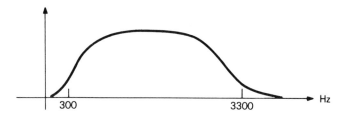

Fig. 12.5: *Frequency transmission characteristic (amplitude only) of a typical telephone channel. Additional phase and nonlinear distortions can degrade transmission even more than amplitude distortions.*

form of frequency modulation and operate at 300 bits/second full duplex. The telephone channel is split in the middle of its frequency band so that each transmission direction has its own half channel.

The rate designation "300 baud" refers to the rate at which digital symbols are applied to the transmission line, which for these modems is identical to the data rate, measured in bits per second. The data rate can be considerably higher. Modems at rates of 1200, 2400, 4800, and 9600 bits/second are commonly used in business applications, and 1200 bits/second has become a *de facto* standard for videotex (Appendix 2), although even at this rate, considered high for consumer applications, information frames with complex graphics can take an excruciatingly long time to fill a television screen. Higher speed and lower cost modems are being made possible by progress in very large scale integrated (VLSI) semiconductor circuits, and consumer electronics will probably include built-in 2400 or even 4800 bits/second modems within a few years.

Fig. 12.6: *Functional structure of a modem. Channel equalization, partially compensating the channel deficiencies and required for higher speed modems, is ordinarily self adaptive.*

Using dialup modems and the ubiquitous telephone network, personal computers and other terminals can quickly establish communications with many data banks and service computers. The host computer providing the data or other services has its own modem, or perhaps a bank of telephone

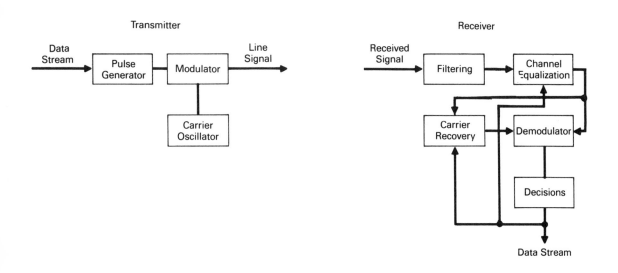

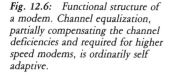

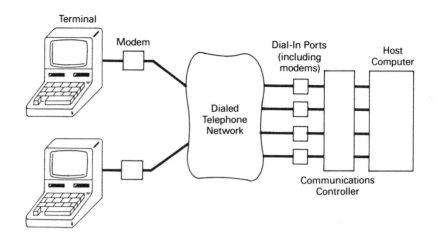

Terminal

Modem

Dial-In Ports
(including
modems)

Host
Computer

Dialed
Telephone
Network

Communications
Controller

Fig. 12.7: Terminal to host computer data communications through the dialed telephone network.

ports each with its own modem, as suggested in Fig. 12.7. Most electronic information and transaction services operate in this way.

For communication with host computers outside of the local calling area, the costs of long-distance calls can make this an expensive accessing method. A telephone channel carrying intermittent data traffic at 300 bits/second, or even 1200 bits/second, is being used very inefficiently. To achieve economies of sharing and provide enhanced services to users, such as speed conversions between different terminals, *packet-switched data networks* have been set up in the United States and other countries. The data transmissions from many terminals are broken into packets of perhaps 1000 bits each, which are individually addressed and sent as they occur through the network. Transmission channels are kept busy more of the time, and the user can be charged less for data transport because only part of a telephone channel is being dedicated to that user.

Access to a packet-switched network is usually by a local telephone call to a "node" of the network (Fig. 12.8). Although telephone carriers such as AT&T, MCI, and GTE are not the only operators of these networks in the United States, packet-switched services are well established in the telephone industry. *Circuit-switched* data services, required on a demand basis by large users with dense point to point traffic, are also provided within the telephone network at a variety of rates.

LADT and ISDN

Even without further enhancements, the universal connectivity of the telephone network, augmented with the cost savings and "added value" services of packet-switched networks accessed by local telephone calls, are difficult to achieve in interactive CATV. But major improvements are being implemented which will enhance these advantages by increasing transmission speed, reducing call setup time, and making most data calls less

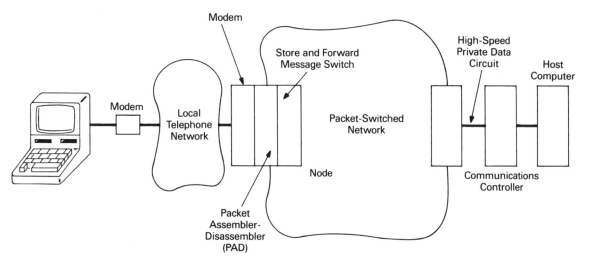

Modem

Store and Forward
Message Switch

High-Speed
Private Data
Circuit

Host
Computer

Modem

Local
Telephone
Network

Packet-Switched
Network

Node

Communications
Controller

Packet
Assembler-
Disassembler
(PAD)

Fig. 12.8: *A dialed data call*
through a packet-switched data
network. The data packets from
many users share the transmission
circuits within the network.

expensive. Videotex communications at (for example) 9600 bits/second instead of 1200 bits/second, allowing video frames to be scanned almost as quickly as pages in a printed magazine, may make videotex more acceptable to the average consumer. This is particularly important in the context of competition with CATV, which has a present advantage over the telephone network only in these areas of quick call setup and fast downloading of information.

The first of these improvements is a packet-switched network operating in a telephone company LATA. (Fig. 12.9). It is often called *local area data transport* (LADT). With the availability of an LADT service, a telephone subscriber has access to two parallel public communication networks: the dialed voice network and the LADT data network. In the "data over voice" access option illustrated (with others) in Fig. 12.9, the data channels are frequency multiplexed over the voice channel in the subscriber terminating equipment, so that voice and data conversations can go on simultaneously over a single line.

The data accessing arrangement on the subscriber's premises allows direct digital data input, removing the need for a modem, although the interfacing equipment utilized by the subscriber incorporates a device which communicates with the local telephone exchange. This direct digital connection eliminates the call setup delays and bandwidth limitations inherent in the system of dialed access to a packet network illustrated in Fig. 12.8.

The first LADT service was initiated in Florida in 1984 to support the Viewtron videotex service offered in the Miami area. Although initial operation was at 1200 bits/second, videotex service at 4800 bits/second and higher rates is envisioned for the future. Service providers, including videotex operators, establish high-speed private line connections to the

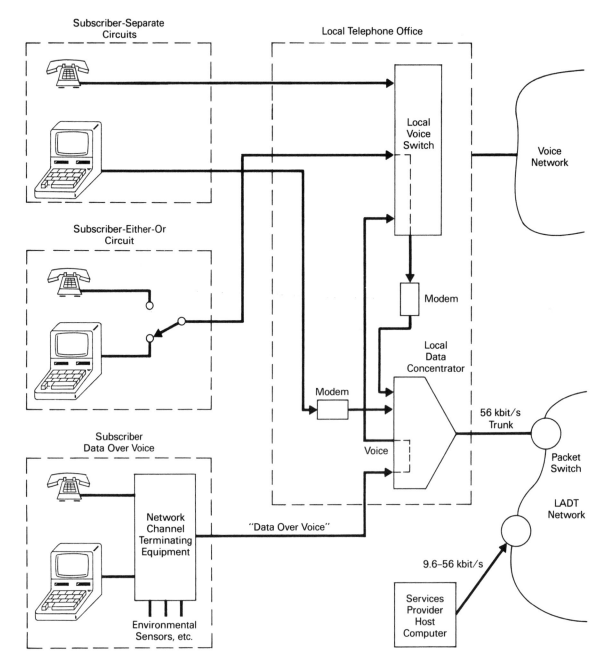

Fig. 12.9: *Local area data transport (LADT), a packet-switched network in the operational area of a local exchange (telephone) company. Data transmission from subscriber to local exchange office is by separate terminating circuit, by either-or use of the telephone terminating circuit, or by "data over voice" in which the frequency spectrum above the voice channel is used for modulated data signals.*

LADT network, or connect through other, wide area packet-switched networks.

Communications deregulation delayed the implementation of LADT networks because of disputes between telephone companies and value-added data network operators over whether this is a basic service within the purview of regulated operating telephone companies, or a value-added service to be offered only by separate, unregulated operators. At first, LADT was restricted to support only of terminal devices with the sophisticated communications protocol used within the network. The federal government later began to relax its prohibitions on the communications protocol conversions which LADT must provide to support inexpensive terminals and connect to other networks, and it is possible that both telephone companies and value-added carriers may eventually offer direct packet network access through the subscriber terminating circuit.

LADT is a forerunner of the integrated services digital network (ISDN), a concept for an end-to-end digital communications network which has been embraced by telephone companies and communications administrations around the world. In addition to digital end-to-end connectivity, the ISDN is supposed to offer a standard subscriber interface usable for a wide variety of services. These include basic circuit-switched voice communications, packet-switched data communications, what appears to be (i.e., "virtual") private line data communications, and specific applications operating at a wide range of data rates. Considerable opportunity is to be given to the subscriber to control the assignment and use of transmission capacity. Specification of the standard interfaces has been part of the work program of the International Consultative Committee on Telephone and Telegraph (CCITT) for some time. Tentative agreement has been reached on the basic 2B + D interface illustrated in Fig. 12.10, among others. A broader view of the concept is contained in Fig. 12.11.

Fig. 12.10: The basic 2B + D subscriber interface proposed for ISDN. The B channels are not necessarily dedicated to the uses shown here.

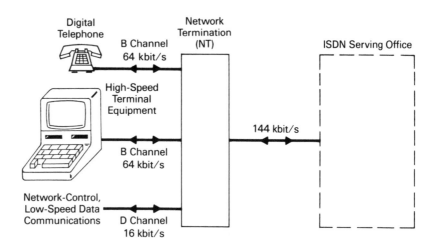

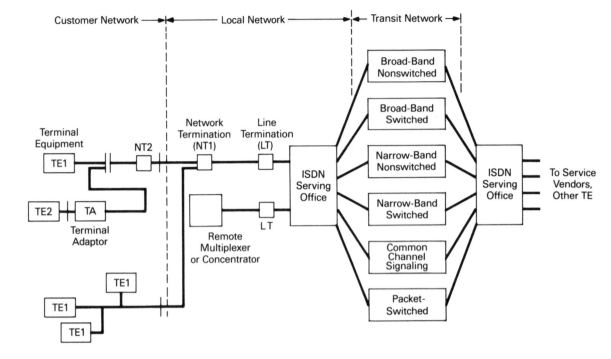

Fig. 12.11: *Architecture and functions of the integrated services digital network. (from Kostas [2].)*

Aside from the integration of voice and data functions and new opportunities for dynamic control of communications networking functions, the most striking feature of the ISDN concept, even in the moderate rate version, is the substantial increase in data communications capacity offered over current telephone line data rates. The term $2B + D$ refers to two 64 kbit/second B channels, and one 16 kbit/second D channel, all full duplex. Thus the subscriber with a basic ISDN interface will have a total digital capacity of 144 kilobits/second full duplex. The B channels are typically allocated to digitized telephony and/or high-rate terminals, while the D channel is used for both signaling (e.g., to set up a call) and lower speed data communications. Even with one of the B channels dedicated to voice communications, data transmission capacity at rates which are very high for consumer services would be simultaneously available.

Higher rate ISDN channels, well into the megabit per second range, are also envisioned. This would serve large commercial customers with bulk transmission needs and customers for videoconferencing and videotelephony. The "primary" interface, with a $23B + D$ configuration, is already well defined.

The ISDN concept is being extended to very high data rates of the order of 45 megabits/second and higher. Light wave transmission and switching systems will be the foundation for this revolutionary communications network of the future, the broad-band ISDN.

The Optical Fiber Telephone Network

Optical fiber as a transmission medium, introduced in Chapter 2 as an element becoming fairly common in cable systems for major trunking applications but not for the subscriber drops, has a similar pattern of use in the telephone network. It is being used very intensively in short- and medium-distance interoffice (between telephone switching offices) carrier systems, and is becoming the terrestrial medium of choice for new long-distance carrier systems. Plans have been made to lay the first transatlantic fiber-optic cable in 1988, with an anticipated capacity of 40,000 telephone channels.

Virtually all of these applications, unlike those of cable operators, are high-rate digital rather than analog. Development work on single-mode fiber-optic systems is looking toward data rates of 1 gigabit/second (one billion bits/second) on a single fiber with a long spacing between regenerative repeaters, sufficient for more than two thousand digitized voice channels or one to two dozen digitized video channels. The reason for the superior quality of digital video transmission is that the same perfect digital data stream, representing one or more video signals, can be delivered to each and every subscriber. In analog systems, such as CATV, subscribers far from the transmitting source receive signals degraded from ingress noise and distorted from passage through many amplifiers, cables, and connectors.

Although the running of fiber-optic cables to individual telephone subscribers has not yet been justified on the bases of cost and demand for wide bandwidth services, there is a feeling in the technical community that this could happen by the end of the century. By then a combination of factors—the obsolescence of the copper telephone wire plant and of the CATV plant installed in the 1970's and early 1980's, the maturing of new electronic communications services, cost and performance improvements of optical communication systems—will make the installation of fiber-optic subscriber lines economically and technically compelling.

The fiber-optic subscriber lines will, of course, be only part of a broad-band integrated services digital network which will support demand access to every conceivable communications service, including normal "cable television" services and their extensions to video programming on demand (Fig. 12.12). A household might have the capability to operate four or five video on demand channels simultaneously, in addition, of course, to several digital stereo music channels, privacy protected digital telephone circuits, and extremely fast information retrieval services. One or two video channels, operating at rates of 130 to 300 megabits/second, could offer extended- or high-definition video signals of outstanding sharpness and color. Pictures might be displayed on the relatively inexpensive projection systems, flat wall screens, and other improved viewing equipment expected in the future. "Telephone offices" might have several thousand video broadcast channels for subscribers to choose from, realizing a service very close to true video on demand.

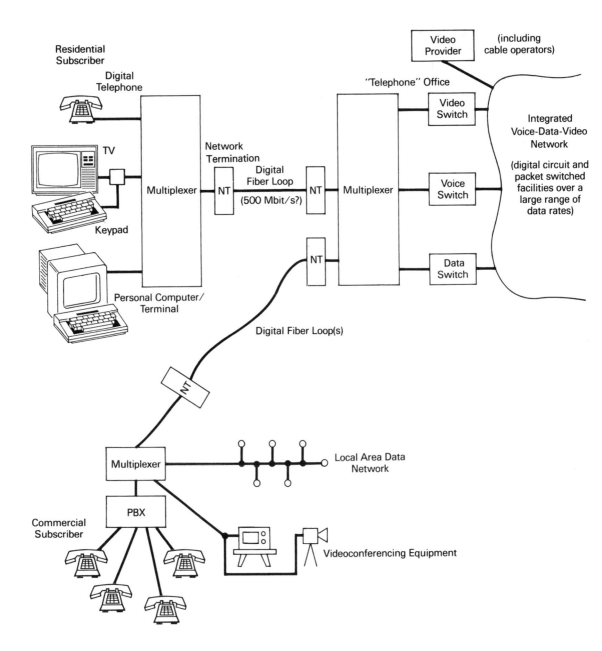

Fig. 12.12: Concept of a switched broad-band network based on fiber-optic digital transmission to the end subscribers.

This is the development which could provide the ultimate competition to cable operators: an alternative digital delivery network so much more flexible and powerful than analog coaxial cable, and with such superior signal quality and reliability that coaxial cable broadcast systems could not survive. Rather than driving cable operators out of business, the chances are that many cable operators would become entertainment packagers and distributors operating as service providers on the optical fiber system, which probably would be, although it does not have to be, owned by local exchange telephone companies. There may be farsighted cable operators who are already planning this transition, and are minimizing their projected investments in cable facilities which could become obsolete before completing their normal service life.

It is, of course, not absolutely certain that this quantum leap in delivery technology will be realized in the consumer market, but the chances are good. If the consumer thirst for better and more useful video, not necessarily more video, which is now being stimulated by VCR's, home video production equipment, and improved television sets, can be channeled into a coherent demand for a full-services broad-band network, that network will be built. Early business demand for enhanced communications will be a large contributing factor. People can argue about when this will happen, but who can doubt that a technical development so appealing to our sensual and information needs will eventually become part of our lives?

For Further Reading

[1] F.T. Andrews, Jr., "ISDN '83," *IEEE Commun. Mag.*, Jan. 1984.
[2] D.J. Kostas, "Transition to ISDN—An overview," *IEEE Commun. Mag.*, Jan. 1984.
[3] S.D. Personick, *Fiber Optics: Technology and Applications.*   New York: Plenum, 1985.
[4] Special Issue of *IEEE Spectrum*, "The future of telecommunications," Nov. 1985.
[5] Special Issue of *IEEE Commun. Mag.*, "Divestiture—two years later," Dec. 1985.

1

Appendix
The Television Signal

This extremely concise description of a color television signal, generally following the articles by Deutsch [1] and Pritchard and Gibson [2], provides background information for discussions in the body of the book. In-depth background material may be found in the reprint volumes by Rzeszewski [5], [6] and information on more recent advances can be found in the April 1985 issue of the PROCEEDINGS OF THE IEEE [9].

Television, like movies, generates the illusion of movement by rapid presentation of a series of still pictures. Three different incompatible color systems, designated NTSC, PAL, and SECAM, are "standard" in different parts of the world. The NTSC (National Television System Committee) color system used in North America and Japan is described here. The system is completely compatible with the older black and white system; color transmissions are displayed in black and white on black and white television sets, and color sets can display black and white broadcasts. The color television signal fits into the same 6 MHz bandwidth as a black and white signal.

Color television is designed to convey to the viewer the characteristics of brightness (relative intensity), hue (color or optical wavelength), and saturation (vividness of a color) which define the human sensation of color vision. Specific steps were taken in the design of the television signal to transmit these characteristics with minimal subjective distortion.

Each displayed "frame" consists of 525 horizontally swept lines, each of which can contain roughly 425 distinct picture elements (pixels). A frame is swept out by an electron beam line by line. To avoid flicker, the frame is actually displayed in two interlaced "fields," each of which contains half of the lines. Thirty frames, corresponding to sixty fields, are displayed each second.

During the vertical blanking interval (VBI) at the end of each field, when the electron beam returns to the top of the screen, 15 or 16 horizontal traces are suppressed or hidden above the visible screen. During this interval, no useful information about the television picture is sent, and the time can be used to transmit data, as in teletext services (see Appendix 2). Each television frame displays only about 490 of the nominal 525 lines.

With 30 frames per second and 525 lines, visible or suppressed, per frame, 15,700 lines are traced per second. Actually, a slightly different rate of 29.97 frames per second, corresponding to 15,734 lines per second, is used to minimize interference between the color subcarrier and 4.5 MHz sound subcarrier. A horizontal sweep generator in the television set, operating at this 15.734 kHz frequency and synchronized with pulses in the incoming signal, controls the tracing of horizontal lines. A vertical sweep generator, also synchronized with the incoming signal, moves the electron beam down the screen and back to the top each 1/59.94 second, corresponding to each field.

The television signal itself, whether broadcast over the air or carried in a cable channel, consists of a "composite video signal" modulated onto a carrier waveform and joined by an FM audio signal. As Fig. A1.1 shows, the composite video signal is produced from three time-varying signals: a brightness or luminance signal Y, and two additional color difference signals I and Q, which are all derived from the red (R), green (G), and blue (B) outputs of a color TV camera. The I signal, representing the orange-cyan colors, has a bandwidth of about 1.5 MHz, and the Q signal, representing the less sharpness-critical green-purple colors, has a bandwidth of about 0.6 MHz. Y, with a wide bandwidth of 4.2 MHz for adequate resolution of pixels, is a particular combination of R, G, and B, which is subjectively a good representation of luminance. The amplitude and phase relationships of the three signals, after they are modulated onto carrier signals, convey the hue and saturation information. A television receiver reverses the transfor-

Fig. A1.1: *Generation of an NTSC composite video signal.*

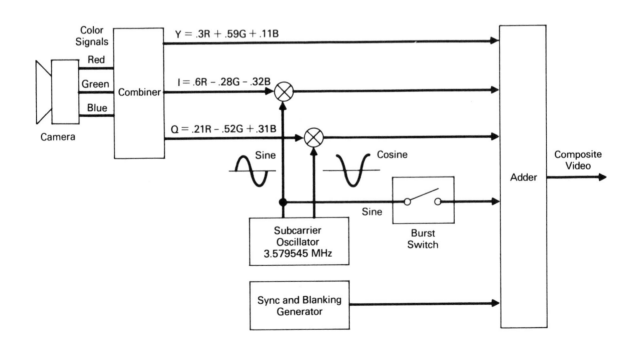

mation to recover the three original color signals, and it also uses the luminance signal directly.

The frequency spectra of the Y, I, and Q signals are not continuous, but instead have a periodic structure resembling a comb, with the major "teeth" spaced at 15.734 kHz intervals, i.e., the horizontal sweep frequency. The I and Q signals, where the letters stand for "in-phase" and "quadrature" respectively, amplitude modulate (with suppressed carrier) sine and cosine waveforms at the *color subcarrier* frequency of 3.579545 MHz, as shown in Fig. A1.1. By using this particular frequency, which is an odd multiple (455) of one-half of the horizontal scanning rate (7.867 kHz), the luminance and chrominance comb frequency spectra *interleave* (Fig. A1.2).

A television receiver with a comb filter, which many modern receivers have, can separate the luminance signal from the modulated I and Q signals by exploiting this interleaving of spectra which at first glance would appear to interfere with one another. This results in a considerable subjective improvement in detail and color separation.

The cosine subcarrier waveform is, of course, the same as the sine waveform except for a 90° phase shift, and if the receiver has one of these sinusoids as a reference it can produce the other and separate the I from the Q modulation. This phase reference information is supplied to the receiver in a *color burst synchronizing* transmission following the horizontal sync pulse at the beginning of each horizontal trace signal (Fig. A1.3).

To create a modulated color television signal (Fig. A1.4), which can be transmitted over the air or by cable, the composite video signal is vestigial sideband modulated onto a carrier waveform, and the audio signal is FM modulated onto another carrier 4.5 MHz higher. Vestigial sideband modulation is normal double-sideband amplitude modulation with one of the frequency sidebands partially suppressed. The modulated video and audio signals are combined, creating a television signal with the 6 MHz frequency spectrum shown in Fig. A1.5.

Fig. A1.2: *Frequency interleaving of luminance and color spectra of NTSC composite video signal. Sketch is illustrative and not to scale.*

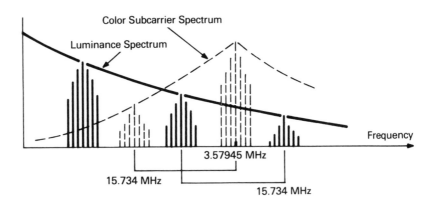

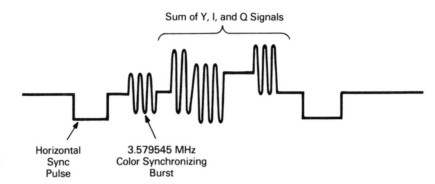

Fig. A1.3: *Composite video signal versus time for a horizontal trace.*

Sum of Y, I, and Q Signals

Horizontal Sync Pulse

3.579545 MHz Color Synchronizing Burst

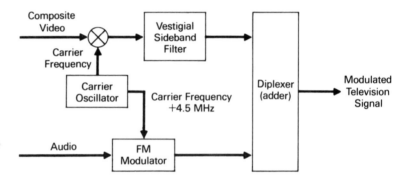

Fig. A1.4: *Generating a television signal from composite video and FM audio.*

Composite Video

Carrier Frequency

Vestigial Sideband Filter

Carrier Oscillator

Carrier Frequency +4.5 MHz

Diplexer (adder)

Modulated Television Signal

Audio

FM Modulator

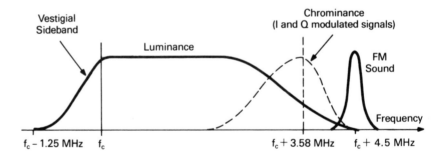

Fig. A1.5: *Spectrum of modulated TV signal.*

Vestigial Sideband

Luminance

Chrominance (I and Q modulated signals)

FM Sound

Frequency

f_c − 1.25 MHz f_c f_c + 3.58 MHz f_c + 4.5 MHz

A composite video signal can be displayed on a standard television monitor, and some consumer television receivers are now being produced in modular form in which the tuner, or radio-frequency receiver, is a separate unit from the video monitor, as suggested in Fig. A1.6. This arrangement, by providing an input to the video monitor which avoids the bandwidth-limiting filters of the tuner, provides a better quality picture from a local program source such as a videocassette player. For computer graphics applications, an even sharper "RGB" monitor accepts red, green, and blue color signals produced by the computer, avoiding the intermediate form, and the compromises, of a composite video signal.

The NTSC color television signal is a compromise among conflicting demands in which the demands for compatibility and minimal bandwidth have won most of the battles. Improvements in signal processing in television sets, including comb filtering, can exploit special characteristics of the television signal in order to improve performance.

Some of these enhancements have only recently becoming feasible with the development of digital integrated circuits and of reasonably priced digital field stores, which hold digitized samples of the luminance and color component signals of a television field. The digital field store facilitates picture-improving filtering and new capabilities such as line interpolation, freeze-frame, and picture-in-picture. Digitization all along the chain of production and distribution, particularly the use of digital component signals in place of the NTSC composite signals, will improve resolution and reduce the degradation caused by repetitive NTSC encoding and decoding. There is still a considerable potential for quality improvement in enhanced NTSC television.

Transmission of component, rather than composite television signals, has

Fig. A1.6: *Television receiving apparatus. A standard television set will have all of these units in one package, but modular systems, with separate units, are becoming available.*

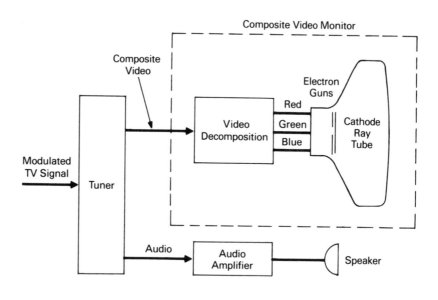

received a great deal of attention. Direct satellite broadcasting, with its special FM noise problems, is one important application. The different components—luminance, color difference, and sound—are time-multiplexed rather than frequency-multiplexed. Different time intervals are allocated in line or frame intervals to the different components. Some or all of the components transmitted in these intervals must be *time compressed*, i.e., transmitted in a fraction of their normal duration, and delayed from their original simultaneous generation times. The receiver must expand, delay and realign them properly.

A family of MAC (multiplexed analog component) systems, in particular C-MAC, D2-MAC, and B-MAC, have been proposed, mainly for DBS applications. All need more bandwidth than an NTSC signal, e.g., 8.4 MHz for C-MAC. They differ mainly in how the digitized sound is realized.

Entirely new color television systems, not constrained by the limitations of a 1940's design, may be developed for high definition television (HDTV). The main attributes of HDTV are a new aspect ratio (of picture width to height) of about 5:3 instead of the present television standard of 4:3, a substantially increased number of scan lines for more vertical resolution, and a greater video bandwidth for more horizontal resolution. There are a number of proposed systems, with the leading contenders the 1125 line, 20 MHz, noncompatible NHK (Japan Broadcasting System) proposal; the 1050 line, 16 MHz, compatible CBS system; and the 8 MHz, noncompatible NHK MUSE system, which delivers a picture that effectively has many lines but requires a bandwidth comparable to that for today's television. For U.S. application, the resolution would be equivalent to about 750 interlaced lines, and could be made compatible with existing TV sets, although the high definition image could be seen only on new HDTV sets.

Television Frequencies

The carrier frequency f_c of Fig. A1.5 can be anything, but standard television channels have been defined in the very high frequency (VHF) and ultra high frequency (UHF) bands. UHF is a poorly chosen name, because television signals are transmitted at still higher frequencies in microwave and satellite services. The large midband gap, between 88 and 174 MHz, is to accommodate FM, air navigation, police, radio amateur, and other services that had claim to those frequencies before television.

The VHF and UHF channels are defined in Table A1.1. All frequencies are in megahertz.

Stereo Sound

Good quality monaural sound is provided for in FCC television standards, which require a sound channel capable of a 50 Hz to 15 kHz frequency

Table A1.1: VHF and UHF channels (frequencies in MHz).

VHF Channel	Channel Frequencies	Video Carrier	UHF Channel	Channel Frequencies	Video Carrier
2	54–60	55.25	41	632.0–638.0	633.25
3	60–66	61.25	42	638.0–644.0	639.25
4	66–72	67.25	43	644.0–650.0	645.25
5	76–82	77.25	44	650.0–656.0	651.25
6	82–88	83.25	45	656.0–662.0	657.25
7	174–180	175.25	46	662.0–668.0	663.25
8	180–186	175.25	47	668.0–674.0	669.25
9	186–192	187.25	48	674.0–680.0	675.25
10	192–198	193.25	49	680.0–686.0	681.25
11	198–204	199.25	50	686.0–692.0	687.25
12	204–210	205.25	51	692.0–698.0	693.25
13	210–216	211.25	52	698.0–704.0	699.25
			53	704.0–710.0	705.25
UHF Channel	Channel Frequencies	Video Carrier	54	710.0–716.0	711.25
			55	716.0–722.0	717.25
			56	722.0–728.0	723.25
14	470.0–476.0	471.25	57	728.0–734.0	729.25
15	476.0–482.0	477.25	58	734.0–740.0	735.25
16	482.0–488.0	483.25	59	740.0–746.0	741.25
17	488.0–494.0	489.25	60	746.0–752.0	747.25
18	494.0–500.0	495.25	61	752.0–758.0	753.25
19	500.0–506.0	501.25	62	758.0–764.0	759.25
20	506.0–512.0	507.25	63	764.0–770.0	765.25
21	512.0–518.0	513.25	64	770.0–776.0	771.25
22	518.0–524.0	519.25	65	776.0–782.0	777.25
23	524.0–530.0	525.25	66	782.0–788.0	783.25
24	530.0–536.0	531.25	67	788.0–794.0	789.25
25	536.0–542.0	537.25	68	794.0–800.0	795.25
26	542.0–548.0	543.25	69	800.0–806.0	801.25
27	548.0–554.0	549.25	70	806.0–812.0	807.25
28	554.0–560.0	555.25	71	812.0–818.0	813.25
29	560.0–566.0	561.25	72	818.0–824.0	819.25
30	566.0–572.0	567.25	73	824.0–830.0	825.25
31	572.0–578.0	573.25	74	830.0–836.0	831.25
32	578.0–584.0	579.25	75	836.0–842.0	837.25
33	584.0–590.0	585.25	76	842.0–848.0	843.25
34	590.0–596.0	591.25	77	848.0–854.0	849.25
35	596.0–602.0	597.25	78	854.0–860.0	855.25
36	602.0–608.0	603.25	79	860.0–866.0	861.25
37	608.0–614.0	609.25	80	866.0–872.0	867.25
38	614.0–620.0	615.25	81	872.0–878.0	873.25
39	620.0–626.0	621.25	82	878.0–884.0	879.25
40	626.0–632.0	627.25	83	884.0–890.0	885.25

response and a 60 dB signal to noise ratio, but TV sets have not been built to provide high fidelity audio. Decades after FM radio converted to stereo, television is beginning to make the change. In April 1984, the FCC approved the Zenith/dbx system for TV signal stereophonic sound, establishing a *de facto* standard which has motivated manufacturers to begin building stereo sound, and much better amplifiers and speakers, into new television sets. The system is compatible with monaural receivers, just as FM stereo broadcasts can be received by monaural FM receivers.

In addition to the stereo channels, a monaural second audio program (SAP) channel is included in the new standard. This channel could be used for a second language or any other desired purpose.

The stereo sound system is similar to, but not exactly the same as, FM stereo. Like an FM stereo signal, the TV stereo signal (Fig. A1.7) consists of a

Fig. A1.7: *The BTSC (Zenith/dbx) television sound system (from Eilers [4]).*

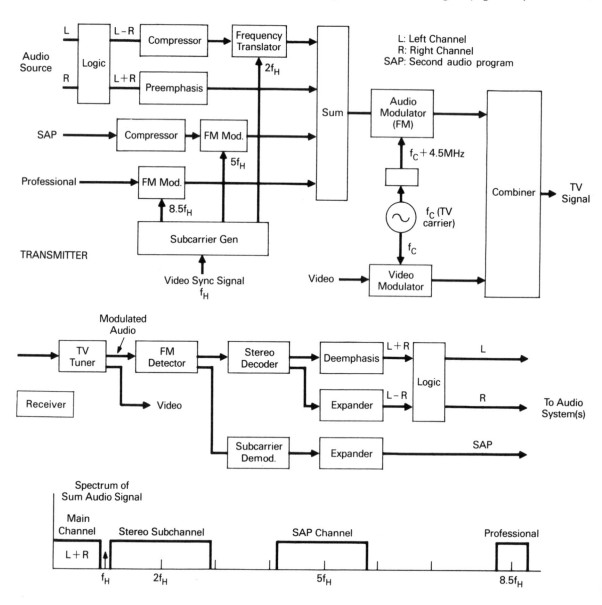

50 Hz–15 kHz "baseband" component which is the combination of the left and right channel signals. This monaural audio signal is FM modulated on the main audio carrier signal, as in ordinary television, and will be received by a monaural system.

An additional signal, representing the difference betwen the left and right channel signals, is frequency translated so that its component frequencies are centered on 31.468 kHz, twice the TV horizontal sweep frequency. Prior to frequency translation, the difference signal is "compressed," i.e., the weaker portions are amplified and the louder portions attenuated, to reduce the damaging influence of the higher frequency noise characteristic of an FM system. The SAP signal, if present, is similarly compressed before being frequency modulated on a subcarrier at five times the horizontal sweep frequency. An additional "professional" audio signal, used for signaling purposes, may also be present, FM modulated on a subcarrier at 8.5 times the horizontal sweep frequency.

In the receiver, the sum, difference, and SAP signals are recovered, the difference and SAP signals by use of the output of the horizontal sweep generator as the demodulation reference carrier. The compressed difference signal, and SAP signal if present, are expanded, and sum and difference logic circuits recover the left and right audio channels. The complex signal processing operations can be carried out in one modern integrated circuit, allowing inexpensive implementations. High-quality stereo sound improves ordinary television sets and encourages the concept of modular home entertainment centers in which the sound amplifiers and speakers are external high-fidelity units.

For Further Reading

[1] S. Deutsch, "Television," *Encyclopedia Americana*, vol. 26, p. 428, 1976.
[2] D. H. Pritchard and J. J. Gibson, "Worldwide color television standards—similarities and differences," *SMPTE J.*, Feb. 1980.
[3] T. Bell, "The new television: Looking behind the tube," *IEEE Spectrum*, Aug. 1984.
[4] C. G. Eilers, "TV multichannel sound: The BTSC system," *Proc. 1984 Int. Conf. on Consumer Electronics*, pp. 234–235.
[5] T. Rszeszewski, *Color Television*. New York: IEEE PRESS, 1983.
[6] ——, *Television Technology Today*. New York: IEEE PRESS, 1985.
[7] E. R. Martin, "HDTV—A DBS perspective," *IEEE J. Select. Areas Commun.*, Jan. 1985.
[8] K. Hayashi, "Research and development of high-definition television in Japan," *SMPTE J.*, vol. 90, Mar. 1981.
[9] A. Netravali and B. Prasada, Eds., Special Issue on Visual Communications, *Proc. IEEE*, Apr. 1985.

2 Appendix
Videotex and Teletext

Videotex and teletext are consumer-oriented data communication systems which deliver information as television screen displays of text, graphics, and video still pictures such as those shown in Fig. A2.1. The videotex or teletext decoder may be either built into a television set (or personal computer or video terminal), or attached as an accessory. Videotex is an *interactive* system in which the user can send as well as receive information, while teletext is a one-way *broadcast* system in which the user selects from a "magazine" of "pages" or "frames."

Videotex is sometimes taken to mean almost any kind of interactive system involving a personal terminal and a computerized database, including the inquiry-response systems used by businesses such as airlines and the text-only information services available to individuals with communicating personal computers. However, it properly refers to one of several systems with specific presentation qualities. The use of multicolored graphics, intended for mass audiences of nonprofessional users, is the most distinguishing characteristic. Videotex also emphasizes consistent communication protocols, coordination among services providers, and simple user procedures in order to make it possible for ordinary people to connect into a wide variety of computer-based services.

The implementation of videotex and teletext systems is much more advanced in Europe than in North America because of the direct interest of European telecommunications administrations in building and operating them. In fact, videotex as a concept originated in the British Post Office and was first commercially realized in England as the "Prestel" service. France has, perhaps, made the greatest investment, with the government providing free terminals and electronic directory service to telephone subscribers. At the beginning of 1985, with at least $280 million spent so far, the French Direction Generale des Telecommunications was giving away terminals at the rate of 100,000 per month, and justifying the cost by the replacement of printed telephone books.

Small-scale commercial systems are in operation in the United States, but a total of only about 5000 terminals had been sold in the U.S. by the end of 1984. The most ambitious effort was the Viewtron system in Florida, a cooperative project of the Knight-Ridder Newspapers and American

Fig. A2.1: A sampling of videotex display frames (Courtesy

Telephone and Telegraph Company (AT&T) which began commercial
operation in October 1983. The high cost of its AT&T "Sceptre" terminal
may have slowed its development, but videotex in general has not
penetrated the U.S. market at the rate which was anticipated a few years
ago. At the beginning of 1985 a well-known consulting company, Link
Resources Corporation, lowered its projection of the number of 1988
subscribers to text-and-graphics videotex systems from 1.9 million to

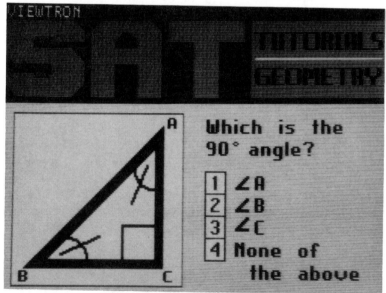

Viewdata Corporation of America, operator of the Viewtron database).

95,000. At the same time, an estimate of 4.2 million was made for 1988 subscribers to text-only systems (not true videotex) compatible with unmodified personal computers and inexpensive video terminals.

Text-only services, using terminals communicating exclusively in ASCII (American Standard Code for Information Interchange) characters, appear to be the wave of the immediate future. It was announced in early 1985 that AT&T and Chemical Bank would begin a joint venture in a text-only service, building on Chemical's existing Pronto home banking service. The joint venture later included AT&T, Chemical Bank, Bank of America, and Time, Inc., and a new, low-cost terminal (Fig. A2.2) was introduced for the consumer market. Other joint ventures, such as Trintex, formed by CBS, Sears and Roebuck, and IBM, and another group composed of RCA, Citicorp, and J.C. Penney, may also initially avoid videotex presentation features in their consumer-oriented systems. True videotex does, however, appear to have growing acceptance in business, government, and institutional applications.

Although videotex and teletext technologies and service concepts did not achieve the success expected by the mid-1980's, they were making progress in the ''high end'' markets, and text-only transactional services, such as home banking and shopping, were being developed by major companies. The mass market, assuming it becomes convinced that electronic information services are worthwhile, may eventually demand the visual impact of videotex and teletext. Public-use terminals are already popping up in airports and hotel lobbies as guides to local places and services.

True videotex may crack the residential market if progress can be made on four fronts:

1) The development of *well-maintained databases of real utility to consumers.*

2) The *speed* of communications and processing so that higher quality visual material can be displayed and fast browsing can be made possible.

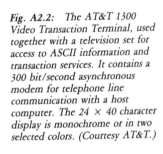

Fig. A2.2: The AT&T 1300 Video Transaction Terminal, used together with a television set for access to ASCII information and transaction services. It contains a 300 bit/second asynchronous modem for telephone line communication with a host computer. The 24 × 40 character display is monochrome or in two selected colors. (Courtesy AT&T.)

3) The *cost and availability* of decoders largely through building cheap decoders into production television sets.

4) The development of improved *user interfaces*, exploiting machine intelligence and the potential of visual displays to make usage sessions convenient, pleasant, and fast.

The transmission requirements for present-day videotex and teletext can be met on many different transmission media. Videotex developed as a telephone-line service and that is the way it is usually delivered, but it is quite feasible with adequate interactive CATV facilities. Similarly, teletext developed as an adjunct to broadcast television, using "free" capacity available during the vertical blanking interval (VBI) of the television frame generation cycle. It can, however, be distributed through any broadcast medium, and in profusion on cable television systems if a whole channel is made available, as described in Chapters 2 and 4. It is important to distinguish, as Chapter 2 does, between teletext delivered as a data stream to a subscriber's decoder, and teletext converted into a television signal and sent, at great cost in channel capacity but saving in subscriber equipment, to ordinary subscriber television sets without decoders.

Private systems for special user groups can be set up either as separate systems, or within public systems using access codes and scrambled transmissions. Some of the most successful applications in public systems have been with business customers such as travel agencies. Complete videotex systems packages are sold by major computer and information systems vendors.

Videotex

Fig. A2.3 illustrates the essential elements of a videotex system. One or more databases of display frames are maintained by the videotex operator or parties connected through "gateways" to the videotex operator. A "frame creation terminal" (Fig. A2.4), perhaps operated by a commercial artist, can be used to create frames. The frames are requested, through two-way communication circuits, by subscribers who view them on television screens or other video displays. Selections of frames are made from a keyboard linked to the videotex decoder.

The subscriber has the capability of searching for and retrieving information frames, perhaps from databases collectively storing millions of frames, and of making requests, reservations, and purchases through services providers participating in the system. Even when the entire database is maintained by the operator, as in some of the European systems, it may be distributed among multiple locations so as to minimize communications costs and computer access delays. For example, a local database in a city might hold the 10,000 frames most frequently requested in that city, a regional database a larger library of perhaps 100,000 frames, and a national database the full complement of several million frames. A request would be routed up the hierarchy until the desired frame were found.

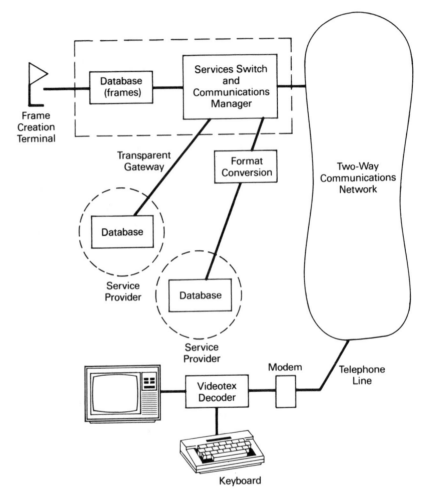

Fig. A2.3: Functional elements of a videotex system. A gateway to a service provider or secondary videotex operator might be either "transparent," with the service provider maintaining a database in videotex format, or non-transparent with a database in some other format which is converted by the primary operator into videotex. The dialed telephone network can potentially be replaced by other interactive networks better suited to data communications, such as local area data transport (Chapter 12) or interactive CATV. The videotex decoder can be realized as a plug-in board for a personal computer instead of a television adjunct.

Fig. A2.4: A frame creation terminal, actually a sophisticated computer-aided graphics design workstation. The rolling "mouse" controls drawing, color selection, and other functions displayed on the screen. (Courtesy AT&T.)

Several alternative methods for accessing a desired frame are provided or at least anticipated in most systems: by *frame number* from a published directory, by *description* such as one or more keywords, by a profile of *attributes* such as the approximate location and price range of a hotel, and by a *tree search* through a hierarchy of "menus." The commonly used but often slow tree search method is illustrated in Fig. A2.5.

The selection of videotex frames in Fig. A2.1 suggests the range of informational and transactional functions which can be supported, but does not do justice to the blinking, animation, translucent and color effects, variable type fonts, and other possibilities within the technology. Properly used, these display attributes can add a great deal of interest to a videotex session and compensate somewhat for the unpleasantness of reading text on a television screen. Other attributes, such as "unprotected fields" in certain frames in which users can enter response information such as credit card numbers, allow users to transact business directly with service providers.

On the other hand, the figure also does not convey the frustrating experience of sitting before a screen and searching for information or waiting for a frame to be "painted." At the data rate of 1200 bits/second commonly used for host-to-subscriber transmissions, a complex graphics

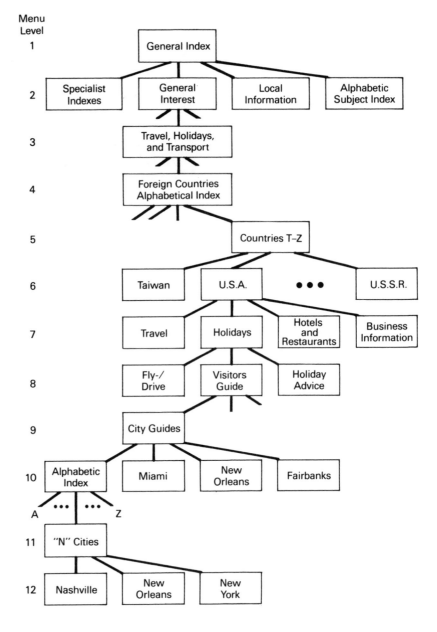

Fig. A2.5: *Illustration of the hierarchy of menus in a tree search for desired information in a videotex database. The branches represent alternative user selections from a menu. This example was drawn from a 1982 search for tourist information on New York City in the British videotex system.*

frame can take almost 10 seconds to complete, and much more time when part of the frame is photographic.

Videotex Presentation Level Protocols

The technical essence of videotex is specified in its presentation level protocol. Almost all computer communications sessions, not only videotex, are regulated by a group of communications protocols which make interchange among different devices possible and effective. These agreements among equipments about how they are to communicate cover everything from definitions of physical connections to the highest level of interworking among end processes or applications. The International Standards Organization has created a hierarchy of seven protocol layers (Table A2.1) called the Open Systems Interconnection Model. The presentation layer is next to the top.

Videotex systems share with other computer communication systems most of the lower level protocols already developed and in use, but because of their special presentation attributes, require their own presentation level protocols. Three mutually incompatible standards for the presentation level protocol exist. In the United States and Canada, the North American Presentation Level Protocol Syntax (NAPLPS), based in large part on the Canadian-developed Telidon system, was adopted in late 1983 by the American National Standards Institute and the Canadian Standards Association. In Europe, the Conference of European Post and Telecommu-

Table A2.1: The seven functionally separate communication protocol layers of the International Standards Organization's Open Systems Interconnection Model.

	Level	Name	Function
(Highest)	7	Application	Interaction with end service or process.
	6	Presentation	Representation of information; codings.
	5	Session	Session connection; control functions for a service or interchange.
	4	Transport	End to end transparent data communications.
	3	Network	Routing and switching; use of transmission resources.
	2	Link	Data transmission via a physical link. Error correction, sequencing and flow control to maintain data integrity.
	1	Physical	Mechanical, electrical and procedural functions, including interface between a terminal and channel communications equipment.

nications Administrations (CEPT) adopted, in 1981, a standard unifying the earlier Prestel and Antiope systems developed in England and France, respectively. Japan has its own standard, the Character and Pattern Telephone Access Information Network (CAPTAIN), oriented toward transmission of Chinese and Japanese characters.

Both NAPLPS and the CEPT standard specify certain *character sets* for coding standard patterns, such as letters and numbers, into transmittable data, and allow for definition of new ones. In NAPLPS, these are a primary set of ASCII characters (an old standard for text transmission), a supplementary set, a mosaic set, a macro set, a dynamically redefinable set (DRCS), and a picture description instruction (PDI) set.

Fig. A2.6 shows the first three sets, which have close counterparts in the CEPT standard. However, the British system incorporated into CEPT is, unlike NAPLPS, *synchronous* in the sense that the location of a character coding in the transmitted data stream associated with a frame corresponds to the location of the character on the screen, which simplifies the decoder design at some cost in flexibility. The question of simple (and therefore cheap) decoders versus more powerful and expensive decoders has been a running dispute between European and American videotex advocates.

The mosaic characters are building blocks for larger patterns, i.e., for graphics. In the standard North American screen layout of 20 lines of 40 character spaces each, figures drawn with these standard building blocks will have a pronounced roughness, as illustrated in the top half of Fig. A2.7. A videotex system based only on characters and mosaics in character spaces is

Fig. A2.6: Standard character sets, including a mosaic set for drawing figures, specified in NAPLPS. The seven bits (b1 through b7) which define a location in one of the tables represent the coding set for that character. A transmission from the host computer will contain an "escape" sequence selecting one of these or other possible sets, followed by codings for characters selected from that set.

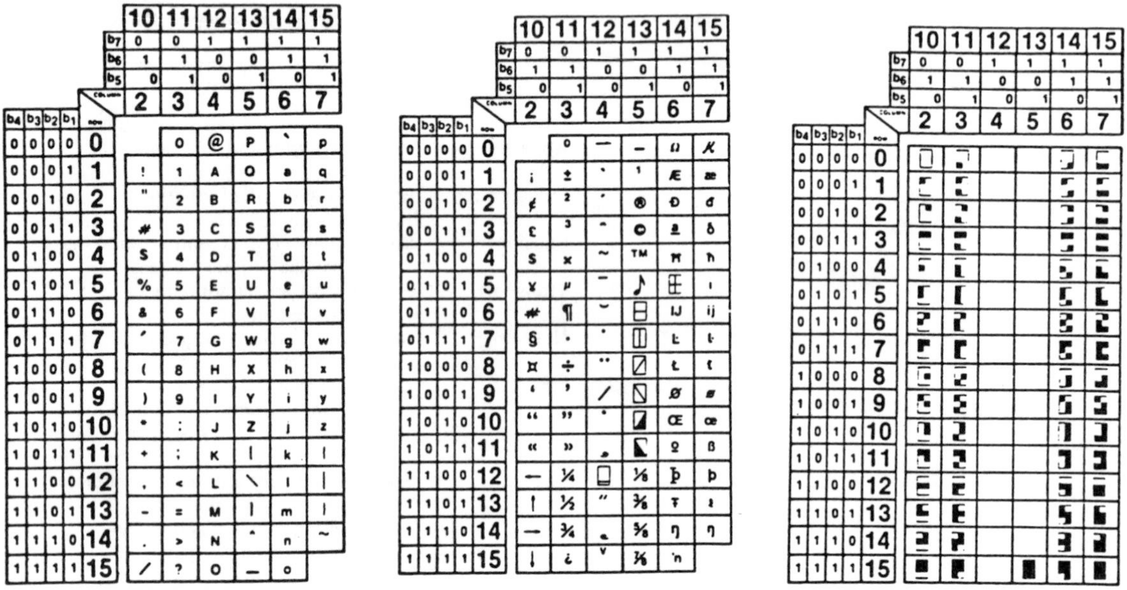

Fig. A2.6 — ASCII (primary) character set

b4 b3 b2 b1	#	2	3	4	5	6	7
0 0 0 0	0		0	@	P	`	p
0 0 0 1	1	!	1	A	Q	a	q
0 0 1 0	2	"	2	B	R	b	r
0 0 1 1	3	#	3	C	S	c	s
0 1 0 0	4	$	4	D	T	d	t
0 1 0 1	5	%	5	E	U	e	u
0 1 1 0	6	&	6	F	V	f	v
0 1 1 1	7	'	7	G	W	g	w
1 0 0 0	8	(	8	H	X	h	x
1 0 0 1	9	)	9	I	Y	i	y
1 0 1 0	10	*	:	J	Z	j	z
1 0 1 1	11	+	;	K	[	k	{
1 1 0 0	12	,	<	L	\	l	|
1 1 0 1	13	–	=	M	]	m	}
1 1 1 0	14	.	>	N	^	n	~
1 1 1 1	15	/	?	O	—	o	

The second (supplementary) and third (mosaic graphic) character sets share the same bit-coding structure, with columns 10–15 (b7 b6 b5 = 001111 / 110011 / 010101) mapped to table columns 2–7 and rows 0–15 indexed by b4 b3 b2 b1.

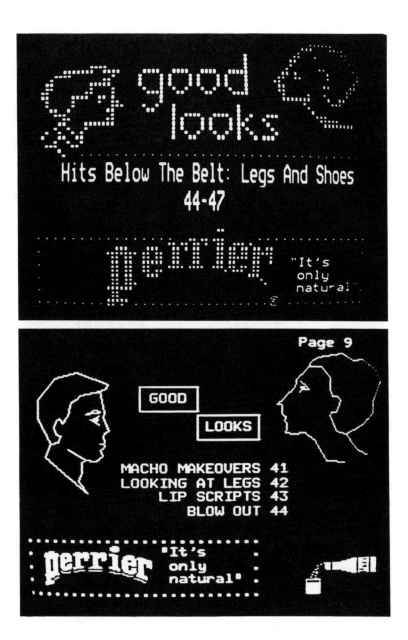

Fig. A2.7: Comparison of resolution in frames drawn using alphamosaic and geometric techniques, respectively. (Courtesy IEEE Spectrum.)

called *alphamosaic*. Some improvement can be obtained by defining additional patterns in the DRCS, which can be sent from host computer to subscriber decoder during the course of a communication session. But superior graphics call for a *geometric* system in which individual picture elements (pixels) can be controlled, rather than settling for a limited number of 6 pixel by 10 pixel mosaic characters.

The scheme used in NAPLPS, but not in the CEPT standard, is one of transmitting picture description instructions (PDI's, Fig. A2.8) which define geometric patterns. This is a highly compressed system for conveying a high-resolution pattern, because information does not have to be sent about each pixel. For example, to draw a line segment, only the start point, end point, and thickness of the line are required. The relatively complex subscriber decoder generates the pixel pattern, or *bit map*, of the video display from these compressed instructions. The complexity lies more in the substantial memory requirement—enough for the color and intensity of 51,200 pixels in the NAPLPS standard screen of 200 pixels high by 256 pixels wide and 16 colors—than in the computational requirement. Character-oriented videotex calls for storage of information on only 800 characters in a 20 × 40 display.

Fig. A2.8: NAPLPS picture description instructions (PDI's) define simple geometric shapes with minimal data.

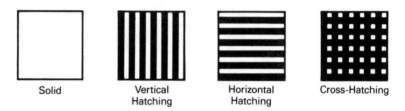

| Solid | Vertical Hatching | Horizontal Hatching | Cross-Hatching |

Filling Textures

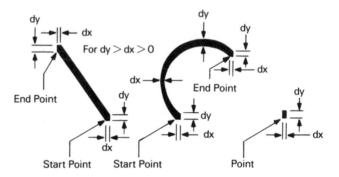

Examples of Basic Elements Describable by End Points
and Thickness in Horizontal and Vertical Directions

NAPLPS PDI's:
• points
• lines
• arcs
• rectangles
• polygons
• solid, dotted, dashed, dot-dashed lines
• thickness and length of line segment

• filling textures
• translucent, highlighting, blinking effects
• incremental point command (pixel by pixel, e.g., for photographs)

Although geometric techniques are powerful aids to achieving high-resolution graphics with minimal transmission of data, they have serious limitations. They are useless for photographs or for graphics which cannot be broken down into line drawings of connected standard elements. For such materials a data intensive, pixel-by-pixel description has to be transmitted, requiring long transmission times at conventional videotex data rates.

NAPLPS and other videotex protocols make provisions for control of features which are not discussed here. These include character colors and gray scale, positionings and rotations of characters, scalable character sizes, and foreground and background colors. The presentation level protocols emphasize standard defaults while offering the systems designer as much flexibility as possible to do more customized and complex frames.

Teletext

It looks the same as videotex, and has a similar presentation level protocol, but teletext is a fundamentally different service. It is a one-way *broadcast* service (Fig. A2.9) in which the user must select from a "magazine" of several hundred frames broadcast over and over again.

The North American Broadcast Teletext Standard (NABTS) developed in the United States incorporates much of NAPLPS, and like NAPLPS is in competition with other standards, particularly the "World System" standard based on British teletext. The British system has a somewhat simpler decoder, analogous to the CEPT videotex system, with transmission synchronized with the horizontal sweep of a television receiver. American broadcasters were experimenting with both systems at the time this was written.

Fig. A2.9: Teletext, a broadcast magazine of information frames, is normally piggybacked on a television signal. The viewer can switch to either normal TV or teletext, or text can be superimposed on the television picture, as in captioning systems.

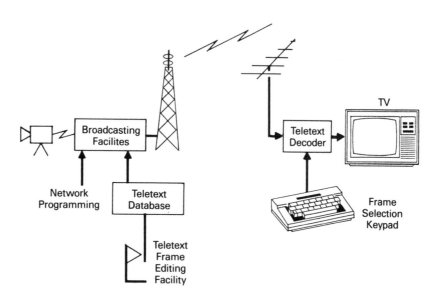

The standard medium for teletext transmission is modulation onto the picture carrier of a television signal during the vertical blanking interval (VBI) following each television field, when no picture information is being sent. This is the time alloted for the picture-tracing electron beam of a television tube to return to the top of the screen. It corresponds to the time required to trace approximately 20 lines of a new frame, although only a few of these lines may actually be used for teletext. One or two of these lines can sometimes be seen as noise at the top of a TV screen. The FCC in 1983 authorized VBI teletext, with none of the rules and regulations, such as "fairness" and equal time requirements, imposed on television broadcasting. A 1985 FCC order opened up the vertical blanking interval for broader classes of ancillary communications services.

There is enough capacity for transmission of a magazine of about 200 teletext frames every 10 seconds. A "frame grabber" built into the teletext decoder permits the user to request a given frame (selected from an index frame) and grab it as it comes by in the transmission cycle. The maximum wait for a requested frame is the interval between transmissions of that frame, e.g., the cycle time of 10 seconds suggested above.

Despite being a broadcast service, teletext can be used for private messaging. Scrambled frames, analogous to telegrams, can be included in teletext magazines, with the teletext decoder of each private message subscriber provided with a personal descrambler. Unauthorized viewers see a scrambled display. This feature is useful in business applications, where a closed user group, as in videotex, can receive information which is not available to others on the system.

Hybrid teletext–videotex systems are also possible, in which commonly requested frames are continually broadcast and others are interactively requested, with the goal of reducing overall transmission requirements. The decoder can be designed so that the user is unaware of how a requested frame is delivered. Other enhancements, including sound, are also possible in the future.

The commercial development of teletext has proceeded slowly in the United States, with CBS's "Extravision" and Taft Broadcasting's "Electra" the prominent developers at the time of writing. Time, Inc., was active earlier, but dropped out in 1984, discouraged over the prospects for low priced (perhaps $150) decoders. Cable-borne teletext has also been pursued, with Group W Cable announcing, in 1984, a full-channel system, "Request Teletext," with 5000 pages of news, weather, sports, retail catalog information, and financial news.

The near-term future of teletext may be brighter than that of videotex, provided there is some consensus on a standard and television set manufacturers are able to build in decoders at a modest incremental cost. Other uses of broadcast media for data broadcasting, such as the commercial VBI data broadcasting which the Public Broadcasting System planned to carry out through its affiliates, may become more significant than mass-audience teletext broadcasting.

For Further Reading

[1] G. D. Ott and J. C. Strand, "Videotex tutorial," *J. Telecommun. Networks*, vol. 2, no. 4, Winter 1983.

[2] *North American Presentation Level Protocol Syntax*, American National Standard X3.110-1983, American National Standards Institute, New York, NY, 1983.

[3] G. Childs, "The European videotex standard," *Comput. Commun.*, vol. 5, pp. 226–233, Oct. 1982.

[4] S. Fedida and R. Malik, *The Viewdata Revolution*. London: Associated Business Press, 1980.

[5] G. Hudson, "Prestel: The basis of an evolving videotex system," *BYTE Magazine*, July 1983.

[6] Interactive picture information systems, *IEEE J. Select Areas Commun.*, vol. SAC-1, no. 2, Feb. 1983.

Glossary

AC (Alternating Current)—An electrical signal with a sinusoidal waveshape, as with the normal 60 Hz power in the United States.

ACCESS CHANNELS—Channels available at no cost to the public and local institutions.

ADDRESSABLE—Responsive only to received data signals identified as intended for that recipient. Often associated with converters.

AD INTERCONNECT—A physical and operational linking of cable systems for the purpose of soliciting and distributing advertising.

ALPHAMOSAIC—A videotex presentation format which displays characters and simple geometric building blocks in character spaces on the display screen.

AML (Amplitude Modulated Link)—A microwave system in which one or more video channels can be modulated onto a carrier waveform by frequency translation.

ANALOG (transmission)—Communication by continuous electrical signals proportional to the original signals.

ASPECT RATIO—Ratio of width to height of a television screen.

AZIMUTH HELICAL SCANNING—An enhancement of the helical scanning tape recording technique in which a relative deflection of the recording heads helps eliminate interference between adjacent recorded tracks.

B CHANNEL—A 64 kbits/second two-way digital channel available through an ISDN subscriber interface.

BANDWIDTH—The breadth (in Hertz) of a band of frequencies available for a communications signal or signals.

BASEBAND—The original unmodulated form of an information signal.

BASIC ISDN INTERFACE—The digital subscriber interface (primarily in the future telephone network) consisting of two 64 kbits/second "B" channels and one 16 kbits/second "D" channel.

BASIC SERVICE (or tier)—The channels and services offered to a cable subscriber for the minimum monthly subscription fee.

BETA (or BETAMAX)—A videotape cassette format.

BIT—A unit of information equal to the information in the choice between two equally likely alternatives.

BPS (bits per second)—A measure of transmission data rate.

BRIDGER—A distribution point in a cable system where an incoming trunk or feeder cable is branched into a number of outgoing cables.

BTSC (Broadcast Television System Committee)—A committee of the Electronic Industries Association whose name has been given to a television stereo sound system.

BYTE—A digital information word consisting of eight bits.

CABLE ACT—The Cable Communications Policy Act of 1984.

CAMCORDER—A combined video camera and miniature videocassette tape recorder.

CARRIER SIGNAL—An electromagnetic waveform which is modulated (varied in some way) with an information signal and thus carries that signal through a transmission medium.

CARS (Cable Antenna Relay Service)—A microwave service, in the 10-13 GHz frequency region, authorized by the FCC for use by cable systems.

CATV—Cable television. An older meaning was "community antenna television."

C-BAND—Frequencies in the 4 and 6 GHz regions used for terrestrial and satellite communications.

CCITT—French initials for International Consultative Committee on Telephone and Telegraph, a standards body under the International Telecommunications Union.

CEPT—Conference of European Post and Telecommunications Administrations. Sometimes refers to communication protocols endorsed by them.

CHARACTER GENERATOR—A device for generation of character displays (text and elementary graphics) on video screens.

CHURN—Subscription turnover, often expressed as the percent of subscribers to a given service cancelling each year.

CLUSTERING—Consolidation of operations of neighboring cable systems. Also used for colocation of satellites.

COAXIAL CABLE—A wide–band electromagnetic transmission medium consisting of an inner conducting core and an outer conducting sheath separated by a nonconducting "dielectric" medium.

COLOCATION (of satellites)—Location of two or more communication satellites close enough in geostationary orbit so that a fixed earth station antenna views them together.

COMPOSITE VIDEO—The baseband color television signal.

COMPULSORY COPYRIGHT—A requirement of federal copyright law on broadcast programmers that cable operators be licensed for a prescribed flat fee to retransmit imported broadcast signals.

CONUS—Continental United States; usually used to describe satellite coverage.

CONVERTER—An electronic device for moving signals from one frequency band to another.

CRT—Cathode ray tube, as used in television sets and data terminals.

CSMA (Carrier Sense Multiple Access)—A random access data communication system in which participating stations listen for other transmissions before beginning their own.

C-SPAN (Cable Satellite Public Affairs Network)—A public affairs program service.

D CHANNEL—A 16 kbits/second data communications and control signaling channel available through an ISDN subscriber interface.

DBS (Direct Broadcast Satellite Service)—Broadcasting from a satellite to an earth station (usually small and low cost) operated by the end user.

DECIBEL (dB)—A measure of power gain or loss, defined as 10 log (P2/P1), where P1 and P2 are input and output powers, respectively.

DEMAND ASSIGNMENT—Allocation to users of communications capacity, e.g., transmission channels, on the basis of current requests.

DIGITAL (transmission)—Communication by a string of discrete-valued pulses, which may result from analog to digital (A/D) conversion of a voice or video signal.

DIRECTIONAL COUPLER—A connection arrangement, usually in a bridge, in which certain directions of signal movement are attenuated much more than others. In cable, the purpose is to isolate interference from subscribers.

DOWNCONVERTER—A device, usually on or near a receiving antenna, to convert the received band of frequencies to a lower band appropriate for cable transmission and further processing.

DOWNSTREAM—The transmission direction from cable headend to subscribers.

DRCS (Dynamically Redefinable Character Set)—A limited collection of videotex or teletext display characters which can be changed during operation.

DROP CABLE—The cable from the tap on a feeder cable to the subscriber's television set.

DUPLEX—Two-way. "Full duplex" means simultaneous two-way, such as separate transmission channels in each direction.

EARTH STATION (Satellite Earth Station)—An antenna, amplifier, and frequency converter installation for reception of signals from communication satellites.

EIA (Electronic Industries Association)—A trade association which encourages standardization.

ENCRYPTION—The transformation of a block or stream of data into another in order to conceal its contents or for some other purpose.

FDM (Frequency-Division Multiplexing)—A technique for placing individual communication signals in contiguous bands so they can be transmitted as one combined signal.

FEEDER CABLE—The cable from which taps are made for the final runs to individual subscribers.

FIELD—A subset of horizontal traces within a television display frame. One of two interlaced fields in an NTSC frame.

FM (Frequency Modulation)—A method of impressing information on a "carrier" signal by varying the carrier signal's frequency.

FOOTPRINT (of a satellite)—The illumination pattern on the Earth's surface of the signal energy from a communications satellite.

FRAME—One completely swept television display.

FRANCHISE—An exclusive license to operate a cable system in a limited area. Sometimes used to refer to the system itself.

FULL–FIELD TELETEXT—Use of a full television channel for teletext transmission.

GB/s (Gigabit per second)—One billion bits per second.

GEOMETRIC GRAPHICS—A videotex display technique in which graphics are developed from instructions, encoded in the transmission data stream, for lines, arcs, and other fundamental geometric shapes.

GEOSTATIONARY ORBIT—An orbital location for a communications satellite which is fixed above a point on the equator.

GHz (Gigahertz)—Billions of Hertz (Hz).

HARMONIC (of a signal component)—A sinusoid whose frequency is a multiple (e.g., 3, for the third harmonic) of the frequency of a sinusoidal signal component.

HBO (Home Box Office)—A satellite-distributed pay program channel.

HDTV (High Definition Television)—A higher resolution television system which requires two to four times the transmission bandwidth of NTSC television.

HEADEND—The primary transmission point in a cable system supplying the hubs and trunk cables.

HELICAL SCANNING—A tape recording technique in which a tape

engages a recording drum in a helical fashion, resulting in many narrow tracks arranged diagonally across the tape.

HRC (Harmonically Related Carrier)—A modulation system for cable headends which harmonically relates the channel carrier frequencies to reduce the effects of harmonic and intermodulation distortions.

HUB—A distribution point in a cable network in which a feed from the headend is branched out into trunk cables.

HYBRID SYSTEM—A system combining cable (or some other wide-band broadcasting medium) with another transmission medium, usually the telephone network, for interactive data communications.

Hz (Hertz)—The basic unit of frequency; cycles per second.

IMPORTED SIGNALS—Television signals, usually of distant broadcast stations, brought to a cable headend for local distribution. Not usually taken to include pay or other channels specifically intended for distant distribution.

INDAX—A commercial system offered, in the past, by Cox Cable, for interactive services on cable systems.

INGRESS—Undesirable electromagnetic noise and interference picked up by a cable, often through poorly maintained connectors and wiring.

INSERTION LOSS—The reduction in signal strength along a cable due to insertion of a device such as a tap. Usually measured in decibels (dB).

INSTITUTIONAL CABLE—That part of an urban cable system, in a separate cable or cables, dedicated to public institutions and other non-residential users.

INTERACTIVE (Bidirectional)—In the cable context, two-way data transmission between subscriber and cable headend.

INTERFERENCE (Cable-generated)—Signals leaking from a cable which interfere with radio communication services.

INTERMODULATION DISTORTION—Self-interference in a cable signal from "mixing" of different channel signals caused by nonlinear imperfections in transmission equipment.

ISDN (Integrated Services Digital Network)—A concept for an end-to-end digital "telephone" network with common user interfaces for all services.

ITFS (Instructional Television Fixed Service)—A microwave band assigned by the FCC for educational purposes.

ITU—International Telecommunications Union, a U.N. agency.

KHz (Kilohertz)—Thousands of Hertz.

K_u–BAND—Microwave frequencies in the 12 GHz band used for business and DBS satellite services.

LADT (Local Area Data Transport)—The regional packet-switched data network operated by a local telephone company.

LATA (Local Access Transport Area)—An operational area of a telephone company.

LEASED CHANNELS—Cable channels offered, for a fee, to outside parties.

LEASED PLANT—Cable plant leased by a cable operator from an alternative owner, such as a telephone company.

LINE EXTENDER—An amplifying bridge installed at the end of a trunk cable.

LNA (Low-Noise Amplifier)—A microwave amplifier, usually close to an antenna, which adds relatively little noise to the amplified signal.

LOCAL ORIGINATION (channels)—Cable channels carrying programming of local origination and interest.

LPTV (Low Power Television)—The class of low-powered ''drop in'' television broadcast stations authorized by the FCC in 1983.

MATV (Master Antenna Television)—A local private television distribution system consisting of broadcast television receiving antennas, amplifiers, and cabling frequency used in apartment buildings.

MDS (Multipoint Distribution Service)—A video broadcasting service utilizing microwave frequencies (2 GHz band) and an omnidirectional terrestrial radiator.

MHz (Megahertz)—Millions of Hertz.

MICROWAVE—An electromagnetic signal at frequencies above about 2 GHz, equivalently expressed as wavelengths less than about 15 centimeters.

MICROWAVE COORDINATION—The process of ensuring that microwave links do not interfere with one another.

MIDSPLIT CABLE—A cable with transmission frequencies split, more or less in the middle, into large upstream and downstream bands.

MMDS (Multichannel Multipoint Distribution Service)—An MDS service utilizing new microwave channels previously dedicated to the Instructional Fixed Television Service.

MODEM (MOdulator-DEModulator)—A device placed between data equipment and an analog communication circuit such as a telephone line, which produces modulated line signals and facilitates data transmission over the analog circuit.

MODULATION—Variation of some parameter of a carrier signal in accordance with an information signal.

MOSAIC GRAPHICS—A videotex display technique in which graphics are built from a limited set of pattern "tiles" occupying a character space on a CRT.

MSO (Multi-System Operator)—A cable company which operates more than one local system.

MULTIMODE FIBER—The optical fiber used with light sources emitting a relatively broad and incoherent set of lightwaves.

MUST-CARRY RULE—The (former) FCC requirement that local broadcast television signals be carried by a cable system.

NABTS (North American Broadcast Teletext Standard)—A teletext standard with many of the presentation features of NAPLPS.

NAPLPS (North American Presentation Level Protocol Syntax)—A videotex standard which makes possible fine color graphics.

NARROW-BAND—A narrow frequency band typically comparable to the 3 KHz bandwidth of a telephone channel.

NARROWCASTING—Programming directed to a smaller than general audience.

NCTA (National Cable Television Association)—An association of cable operators.

NSCA (National Satellite Cable Association)—An association of private cable (SMATV) operators.

NTIA (National Telecommunications and Information Agency)—An advisory body within the U.S. Department of Commerce.

NTSC (National Television System Committee)—Usually used to refer to the U.S. color television system.

OCC (Other Communications Carrier)—An alternative to AT&T.

OPTICAL FIBER—A thin glass cable for transmission of light waves.

OVERBUILD—A second distribution system, e.g., a cable network, in an area already served by a similar distribution system.

PACKET COMMUNICATIONS—A data communications system in which messages are broken into small packets which are individually transported through the network.

PAY–PER–VIEW—Payment by individual program or by viewing time.

PAY TV—Programming paid for over and above a basic subscription rate (if any).

PIXEL (Picture Element)—The smallest resolvable spot in a video display.

POLE ATTACHMENT—Rules and fees for CATV leasing of space on utility poles.

POLLING—A communications procedure in which terminals are interrogated or "polled" from a central controller.

PROTOCOL—An agreement on formats and procedures between elements of a system to facilitate their interaction.

QUBE—An early interactive cable system offered by Warner Amex Cable Communications.

RANDOM ACCESS—A data communications system in which participants enter and leave at their own discretion.

RARC (Regional Administrative Radio Conference)—A regional (e.g., Western Hemisphere) international conference to assign frequencies and other transmission privileges.

RASTER—The pattern of scanned picture elements in a television frame.

SAP (Second Audio Program)—An additional monaural channel provided in the BTSC television stereo sound system.

SATELLITE (Communications)—An orbiting vehicle carrying microwave retransmission stations (transponders).

SCRAMBLING—Distortion or transformation of a signal to conceal its contents or grossly degrade received quality for unauthorized recipients.

SCRAMBLING CONSORTIUM OR COOP—An association of satellite programmers, cable operators, or others for unified control of descramblers in home earth stations.

SINGLE-MODE FIBER—An optical fiber which, by transmitting only a single electromagnetic mode, limits dispersion and supports high data rates.

SMATV (Satellite Master Antenna Television)—A master antenna television (MATV) system incorporating a satellite receiving station.

SPACE (Society for Private and Commercial Earth Stations)—A trade association representing manufacturers and distributors of earth stations.

STRAND—The steel cable, supported by utility poles, from which coaxial cable is hung.

STRAND MAP—A map of utility poles and wiring for use in planning cable wiring.

STV (Subscription Television)—Pay channel delivery through a scrambled television broadcast signal.

SUBCARRIER—A subsidiary carrier waveform, itself modulated by an information signal, which is, together with other signals, modulated on the main carrier waveform.

SUBSCRIBER HUB—A small, remote-controlled cable switching center serving subscribers.

SUPERSTATION—A television broadcast station relayed by a communication satellite to distant locations.

SUPERTRUNK—A high performance coaxial or optical fiber link often used to connect a headend with major distribution hubs.

SWITCHED STAR—A cable system architecture with switching hubs supplying subscribers through individual private cables.

TAP VALUE—The reduction factor in signal strength from feeder cable to subscriber drop cable introduced by a tap. Usually measured in decibels (dB).

TARIFF—A pricing schedule for regulated communications services approved by the regulatory authorities.

TDM (Time Division Multiplexing)—Time-sharing of a transmission channel by assigning each user a dedicated segment of each transmission cycle.

TELETEXT—Broadcast text and graphics by data transmission in television channels. Frequently sent "piggybacked" on a television signal in the vertical blanking interval.

TIER—A group of channels or services subscribed to as a package.

TRANSPONDER—A retransmission station carried aboard a communication satellite.

TRAP—A filter for either passing or excluding a band of frequencies, usually used to control the delivery of pay television.

TRUNK CABLE—A primary distribution cable carrying signals from the cable headend to distribution hubs.

TVRO—Television Receive-Only satellite earth station.

UPSTREAM—The transmission direction from subscriber to cable headend.

UHF (Ultrahigh Frequency)—Microwave frequencies in the 400–800 MHz range used for television broadcasting, in particular channels 14 and above.

VBI (Vertical Blanking Interval)—That portion of each frame time of a composite television signal during which the scanning beam returns to the top of the screen.

VCR—Videocassette recorder.

VHS (Video Home System)—One of the cassette formats for videotape.

VIDEOTEX—Interactive text and graphics transmission, usually utilizing the telephone network but possible on any interactive data communications system of adequate capacity.

VHF (Very High Frequency)—Microwave frequencies in the 50–400 MHz range used for television broadcasting, in particular channels 2–13.

WARC (World Administrative Radio Conference)—A series of conferences to achieve worldwide agreement on frequency allocations and other compatibility questions for all classes of broadcasting.

WIDE-BAND—A band of frequencies typically comparable to the bandwidth (6 MHz) of a television signal.

Index

Society for Private and Commercial Earth
	Stations 172
software distribution
	(see microcomputer software)
Sony Corporation 185, 191–192
Southwestern Cable Company 2
SPACE
	(see Society for Private and Commercial
	Earth Stations)
special user groups, videotex 229
STC
	(see Satellite Television Corporation)
Storer Cable Communications 8
strand map 21
subscriber-switched optical fiber 139
subscriptions, CATV 7
subscription television 137, 147–155
superstation 69
Supreme Court 109
switched star 17
syndicated program exclusivity 120

tap 40–41
TCI
	(see Tele-Communications, Inc.)
Tele-Communications, Inc. 8, 11
TeleFirst 138
telephone companies 13, 100–102, 112–
	114, 123, 126, 143–145
telephone network 199–213
Teleprompter 2
teletext 50, 95–100, 225–239
television broadcasting 138–139
television signal 215–223
Telidon videotex systems 233
text broadcasting 50
text channels 70
theft of services 5, 127, 151, 158
Time, Inc. 2, 8, 10
Times Mirror Cable 8
transponders 28, 30–31, 170, 173, 174,
	178–181
trap 46, 48–49
tree and branch 17
Trintex videotex service 228
trunk cable 18, 37
Turner Broadcasting 172
TVRO station 25

two-way capability
	(see interactive systems)

U-Matic tape format 191
United States Communications, Inc. 161,
	163
United States Satellite Broadcasting, Inc.
	170, 175
United Satellite Communications, Inc. 170,
	173–174
USA Network 72

VBI
	(see vertical blanking interval)
vertical blanking interval 50, 215, 229, 238
VHS
	(see Video Home System)
Viacom Cable 8
videocassette 139, 185–197
videocassette recorders 139, 185–197
video conferencing 102–103
Video Home System 186, 191–193
video on demand 94, 139, 211–212
video programming 132, 134–141
videotex 95–100, 225–239
Viewdata Corporation of America 226–227
Viewtron videotex service 207, 225–226

Walson, John 1, 3
WARC
	(see World Administrative Radio Confer-
	ence)
Warner Amex Cable Communications 6, 8,
	11, 87–89
Wechsler, Steve 133
wide-band interactive services 142–144
wide-band system 18
wiring costs 6
Wirth, Representative Timothy 113, 128
Wisconsin Bell Telephone Company 13
World Administrative Radio Conference
	176
World System teletext 237

ZView 60

About The Author

STEPHEN WEINSTEIN is a communications engineer with interests in services and social impacts as well as technology. His involvement with CATV, videotex, and other consumer-oriented electronic media developed while he was employed by the American Express Company from 1979 to 1984 as an advisor on technology questions. He previously conducted research in telephone-line data communications at Bell Laboratories. Since September 1984 he has worked for Bell Communications Research, the jointly owned research organization of the Bell Operating Companies, on personalized communications in the telephone exchange network. Dr. Weinstein received his degrees in electrical engineering from M.I.T. (B.S., 1960), the University of Michigan (M.S., 1962) and the University of California at Berkeley (Ph.D., 1966). He is a Fellow of the IEEE and Editor-in-Chief of IEEE Communications Magazine, a publication of the IEEE Communications Society, and served several years as Editor-in-Chief of the IEEE Press.

Dr. Weinstein lives with his wife Judith, in Summit, New Jersey.